THE FUTURE

A DEEP DIVE INTO
DIGITAL TRANSFORMATION AND FUTUROLOGY

WRITTEN BY
NATHAN SMITH

 CULTIVATION

ABOUT THE AUTHOR

Nathan Smith is a digital transformation expert and founder of Cultivation, a software platform that provides businesses with a comprehensive suite of digital transformation tools. Based in Bridgeport, Connecticut, Nathan has been helping entrepreneurs and small businesses grow and survive in highly competitive sectors by providing them with the right internal and external digital transformation strategies. His interest in this field developed from his work in various capacities at different organizations, including the Connecticut Department of Transportation, Hartford Healthcare, and UConn Athletics.

With a vision to help small businesses gain visibility and exposure in the digital world, Nathan established Cultivation in 2017 at the tender age of 18. Since then, the company has evolved into a digital transformation software platform, offering businesses a range of digital tools to help them succeed in today's globalized economy. With Cultivation, businesses can streamline their operations, automate their workflows, and transform their customer experiences. The platform's powerful analytics tools also help businesses to better understand their customers and make data-driven decisions.

Cultivation has become a leader in the digital transformation space, with clients and users in multiple industries across the world. Nathan and his team are constantly innovating, pushing the boundaries of what is possible with digital transformation. They believe that every business deserves to succeed in today's digital age and are committed to helping their clients achieve their goals through the power of technology.

TABLE OF CONTENTS

CHAPTER ONE: INTRODUCTION

Welcome to our book on digital transformation, where we will explore the impact of digital technology on our lives and its importance in modern society and business. In today's fast-paced world, digital transformation has become a buzzword that you may have heard about, but not fully understood. This book aims to provide you with a comprehensive understanding of what digital transformation is and how it has changed our lives.

Firstly, let's define digital transformation. Digital transformation is the integration of digital technology into all areas of a business or society, resulting in fundamental changes in how organizations operate and how people live their lives. It encompasses the adoption of new technologies, processes, and cultural changes to achieve better outcomes, such as increased efficiency, innovation, and customer satisfaction.

Digital technology has significantly changed the way operate as humans, from the way we communicate, shop, learn, work and even entertain ourselves. The emergence of digital technologies, such as smartphones, tablets, and the internet, has revolutionized the way we interact with the world around us. We now have access to an infinite amount of information, products, and services at our fingertips. We can connect with people all over the world, access education and job opportunities, and make transactions with just a few clicks.

Digital technology has also transformed the business landscape and as we can see, companies that fail to embrace digital transformation risk falling behind their competitors and losing market share. The use of digital technology in business has led to increased productivity, enhanced customer experiences, and the creation of new revenue streams. Companies that have successfully implemented digital transformation have streamlined their processes, reduced costs, and expanded their customer base.

The importance of digital transformation cannot be overstated. In today's hyper-competitive business environment, companies that fail to adapt to changing market conditions risk becoming irrelevant. Digital transformation is a crucial tool for businesses to stay ahead of the curve, meet customer demands, and create new opportunities for growth. It allows organizations to be more agile, innovative, and responsive to changing market trends.

Digital transformation is essential for modern society. It has the potential to drive positive social change by creating new opportunities for education, employment, and economic growth. Digital technology has already transformed the way we learn, with online courses and e-learning platforms providing access to education for people around the world. It has also created new job opportunities, particularly in the tech sector, and has the potential to boost economic growth by enabling businesses to operate more efficiently.

However, digital transformation is not without its challenges. The rapid pace of technological change means that businesses must constantly adapt to new technologies and market conditions. This can be costly and time-consuming, particularly for smaller businesses. There are also concerns around the impact of digital technology on jobs and society as a whole. The increased use of digital technology has led to concerns around privacy, security, and the impact on mental health. As more tasks become automated, there is a risk of job displacement and increased inequality. It is therefore important that digital transformation is implemented in a way that benefits society as a whole, rather than just a small elite.

Now, before we dive deeper into the world of digital transformation, let's take a brief historical overview of how technology has evolved over time and how it has impacted society and business. We will expand upon the history in the second chapter of this book, however it is good to have this brief knowledge beforehand.

The history of digital transformation can be traced back to the mid-20th century when the first computers were developed. These early computers were massive machines that occupied entire rooms and were used primarily for scientific and military purposes. However, as technology advanced, computers became smaller, cheaper, and more accessible to the general public. The invention of the microprocessor in the 1970s paved the way for the development of personal computers, which revolutionized the way people work and communicate.

The 1980s and 1990s saw the emergence of the internet and the World Wide Web, which transformed the way we access and share information. This led to the development of e-commerce, social media, and online marketplaces, creating new opportunities for businesses and consumers alike. The rise of mobile devices, such as smartphones and tablets, in the early 2000s further transformed the way we interact with the world around us.

The impact of technology on society and business is not a new phenomenon. Throughout history, technological advancements have transformed the way we live, work, and interact with each other. From the invention of the printing press in the 15th century, which enabled the mass dissemination of knowledge, to the development of the steam engine in the 18th century, which transformed the manufacturing industry, technology has always played a crucial role in shaping society and business.

As we have seen, digital technology has transformed every aspect of our lives and has become a fundamental part of modern society and business. However, as with any major technological advancement, there are both benefits and drawbacks to digital technology.

It is clear that digital technology will continue to play a crucial role in shaping modern society and business. From the rise of e-commerce to the development of artificial intelligence, digital technology has created new opportunities and challenges that businesses and individuals must navigate.

In this e-book, we will explore the world of digital transformation and its impact on modern society and business. We will delve into the benefits and drawbacks of digital technology, and examine the ways in which it has transformed the way we live, work, and interact with each other. We will also explore the role of digital technology in shaping the future of business, and the potential opportunities and challenges that lie ahead.

By the end of this e-book, you will have a thorough understanding of digital transformation and its importance in modern society and business. You will have the knowledge and tools necessary to embark on your own digital transformation journey, whether you are an entrepreneur, a business owner, or an individual looking to stay ahead of the curve. So, buckle up, get ready and join me on this journey as we explore the exciting world of digital transformation and discover what it means for the future of our society and businesses.

CHAPTER TWO: A LOOK BACK IN TIME

In the early days of computing, the idea of a machine that could process information was nothing more than a dream. However, with the invention of the first computers, that dream became a reality, and the world was never the same again.

The first computers were massive machines that took up entire rooms. They were incredibly expensive, and only a few organizations could afford them. However, their impact was immense. They allowed for the processing of large amounts of data in a fraction of the time that it would take a human to do the same task.

One of the earliest computers was the Harvard Mark I. It was created in 1944 by Howard Aiken and Grace Hopper. It was a massive machine that was over 50 feet long and 8 feet tall. It used punched cards to input data and had a memory capacity of 72 words. While the Harvard Mark I was not a commercial success, it paved the way for future computers.

The next big leap forward in computing was the invention of the ENIAC, which stood for Electronic Numerical Integrator and Computer. It was created in 1945 by John Mauchly and J. Presper Eckert. It was the first electronic computer and used vacuum tubes to process information. The ENIAC was a game-changer, and it was able to perform complex calculations in a fraction of the time it would take a human to do the same task. It was also much smaller than the Harvard Mark I, making it easier to use and maintain.

As computers became more powerful and more accessible, they began to have an impact on the world. One of the most significant impacts was in the field of science. Computers allowed scientists to process massive amounts of data and perform complex calculations. This allowed for breakthroughs in fields such as genetics, physics, and astronomy.

Computers also had an impact on the business world. They allowed for the automation of tasks that were previously done by humans, such as accounting and payroll. This freed up workers to focus on other tasks and increased productivity.

As computers became more powerful, they needed to be programmed to perform specific tasks. This led to the development of computer programming languages.

The first programming languages were machine language and assembly language. Machine language is the language that computers understand, and it consists of binary code (0s and 1s). Assembly language is a low-level language that is easier for humans to understand and write than machine language. However, it still requires a deep understanding of computer architecture and is not easy to use.

The first high-level programming language was Fortran. It was created in the 1950s by IBM and was designed to be easier to use than assembly language. It allowed programmers to write code in a more natural language, such as English, and the computer would translate it into machine language.

Another significant programming language was COBOL. It was created in the late 1950s and was designed for business applications. It allowed for the automation of tasks such as accounting and payroll, and it is still used today in some legacy systems.

As computers became more powerful, programming languages became more sophisticated. One of the most popular programming languages today is Java. It was created in the 1990s and is used for everything from mobile app development to building enterprise applications.

The early days of computing were an exciting time. The invention of the first computers changed the world, and the development of computer programming languages allowed for even greater advances. Today, computers are an essential part of our lives, and it's hard to imagine a world without them.

While the invention of the first computers marked a significant milestone in the history of computing, it was the development of the internet that would truly revolutionize the world. The internet is a global network of interconnected computers that allows for the sharing of information and communication between people and devices all over the world.

The idea of a global network of computers was first proposed in the 1960s by J.C.R. Licklider, a computer scientist at MIT. Licklider believed that computers could be used to enhance human communication and creativity, and he envisioned a system where people could access information from anywhere in the world. His ideas were put into practice in the 1970s when the US Department of Defense developed ARPANET, which stood for Advanced Research Projects Agency Network, a network of computers designed to allow scientists and researchers to share information.

Over the next few decades, the internet grew rapidly. By the early 1990s, the internet was already being used by millions of people around the world. However, it wasn't until the creation of the World Wide Web that the internet truly exploded in popularity.

The World Wide Web, or simply the web, is a system of interconnected documents and resources that can be accessed over the internet. It was created in 1989 by Tim Berners-Lee, a British computer scientist who was working at CERN, the European Organization for Nuclear Research.

Berners-Lee's idea was to create a system where information could be easily shared and accessed by anyone, regardless of their location or the type of computer they were using. He developed a set of protocols and standards that would allow computers to communicate with each other over the internet, and he created the first web browser and web server.

The web quickly caught on, and by the mid-1990s, it was being used by millions of people around the world. The web made it possible to access a vast amount of information with just a few clicks of a mouse, and it transformed the way we think about communication and information sharing.

One of the most significant impacts of the internet and the web has been on e-commerce and online communication. E-commerce refers to the buying and selling of goods and services over the internet, while online communication refers to the ways in which people communicate with each other online, such as email, instant messaging, and social media.

E-commerce has grown rapidly since the early days of the internet. Today, people can buy almost anything online, from groceries and clothing to electronics and furniture. E-commerce has also opened up new opportunities for entrepreneurs and small business owners, allowing them to reach customers all over the world and compete with larger companies.

Online communication has also transformed the way we interact with each other. With email, instant messaging, and social media, it's easier than ever to stay in touch with friends and family, no matter where they are in the world. Social media has also become a powerful tool for businesses and organizations, allowing them to connect with customers and supporters in new and meaningful ways.

The early days of computing were marked by innovation and experimentation. From the invention of the first computers to the development of computer programming languages, computing technology continued to evolve and improve over the decades. However, it was the creation of the internet and the World Wide Web that truly revolutionized the world. The internet and the web have transformed the way we think about communication, information sharing, and commerce, and they continue to shape our world in new and exciting ways. As we look to the future, it's clear that computing technology will continue to play a significant role in shaping our world and driving innovation in a wide range of fields.

As we can see, the development of mobile technology has also been a significant part of the history of computing. Mobile technology, as we know it today, refers to the technology that allows us to access the internet, use applications, and make phone calls, all from a handheld device. The first mobile devices were bulky, expensive, and had limited capabilities, but they paved the way for the smartphones we use today.

The first mobile devices were developed in the 1970s and were known as car phones. They were large and heavy, and were only used in cars because they required a large power source. However, these early devices laid the foundation for what was to come.

In the 1980s, mobile phones became smaller and more affordable, making them accessible to more people. The first mobile phone with a flip design is considered to be the Motorola StarTAC, which was introduced in 1996. The StarTAC was a revolutionary phone that was significantly smaller and more portable than previous mobile phones, thanks in part to its flip design.

The StarTAC's flip design allowed it to fold in half, making it more compact and easier to carry in a pocket or purse. The flip cover also protected the keypad and screen from damage when the phone was not in use.

The StarTAC was also the first mobile phone to feature a vibrating alert, which was a significant improvement over the loud and sometimes disruptive ringtones that had been used in previous phones. The phone's battery life was also impressive for its time, with the ability to provide up to 180 minutes of talk time or 72 hours of standby time.

The StarTAC was a commercial success, with over 60 million units sold worldwide. Its flip design and compact size set a new standard for mobile phones, and it paved the way for future designs that focused on portability and convenience.

In 2007, Apple revolutionized the mobile phone industry with the introduction of the iPhone. The iPhone was the first smartphone that combined a touch screen with internet connectivity and a wide range of applications. It was a game-changer, and other companies quickly followed suit with their own versions of smartphones.

Smartphones have had a significant impact on society. They have changed the way we communicate, work, and consume media. They have also made it possible to access information and services from anywhere at any time..

One of the most significant impacts of smartphones has been the rise of mobile applications, or apps. Apps are software programs designed specifically for mobile devices. They are easy to use, and they provide a wide range of functions, from social media to gaming to productivity tools.

Apps have had a profound impact on society and business. They have made it possible for businesses to reach customers in new and innovative ways. For example, e-commerce businesses can now create apps that allow customers to shop directly from their phones, and social media platforms can use apps to offer personalized experiences to users.

Apps have also had a significant impact on society. They have changed the way we interact with each other, and they have made it easier for us to access information and services. For example, health apps have made it possible for people to monitor their fitness and health, while educational apps have made it easier for students to access learning materials.

The development of mobile technology has been a fascinating journey. From the first car phones to the smartphones we use today, mobile technology has changed the way we live, work, and communicate. Smartphones and mobile apps have made our lives more convenient, but they have also raised concerns about privacy and the effects of constant connectivity.

As mobile technology continues to evolve, it will be interesting to see what new innovations will emerge. Will we see new types of mobile devices? Will mobile apps become even more integrated into our daily lives? Only time will tell, but one thing is certain – mobile technology will continue to shape our world in new and exciting ways.

Now social media has become an integral part of our daily lives, with millions of people worldwide using it to connect with friends, family, and even strangers. The emergence of social media can be traced back to the early days of the internet, when online bulletin boards and chat rooms were popular among computer enthusiasts.

The early days of chat rooms can be traced back to the 1970s and 1980s, when online bulletin board systems (BBS) were popular. BBS allowed users to dial in using a modem and exchange messages with other users on the same BBS. While not real-time chat, BBS systems did allow for asynchronous conversations that could span days or weeks.

In the early 1990s, with the rise of the internet, chat rooms began to emerge. The first chat rooms were text-based, and users would enter a virtual room and communicate with other users in real-time. These chat rooms were often focused on specific topics, such as technology or music, and they were typically hosted on websites or through internet service providers.

One of the earliest and most popular chat rooms was called "The Palace," which was launched in 1995. The Palace was a graphical chat room that allowed users to create avatars and move around a virtual world while chatting with other users. It was popular among teenagers and young adults, and it was known for its lively and sometimes raucous conversations.

Other popular chat rooms in the early days of the internet included AOL chat rooms, which were available to users of America Online, and IRC (Internet Relay Chat), which allowed users to join chat rooms on a variety of topics and interests.

However, it wasn't until the creation of popular social media platforms such as Facebook, Twitter, and Instagram that social media truly exploded in popularity. These platforms allowed people to connect with each other in new and innovative ways, sharing photos, videos, and thoughts in real-time.

Facebook, created in 2004 by Mark Zuckerberg, quickly became the most popular social media platform in the world, with billions of active users. It allowed people to create personal profiles, connect with friends and family, and share photos, videos, and messages. Facebook also created a new avenue for businesses to reach potential customers, with targeted advertising and company pages.

Twitter, created in 2006, allowed people to share short messages, or "tweets," with a maximum length of 280 characters. It quickly became a popular platform for celebrities, politicians, and journalists to share their thoughts and opinions with a global audience. Twitter also played a major role in social and political movements, with hashtags such as #BlackLivesMatter and #MeToo becoming powerful symbols of social change.

Instagram, created in 2010, was focused on sharing photos and videos. It quickly gained popularity, particularly among younger users, and was acquired by Facebook in 2012. Instagram has become a major platform for influencers and businesses to market their products and services, with sponsored posts and paid partnerships becoming common.

The impact of social media on communication has been immense, with many people now preferring to communicate through social media rather than traditional means such as phone calls or emails. Social media has also created new opportunities for businesses to reach potential customers, with targeted advertising and influencer marketing becoming increasingly popular.

However, social media has also had its downsides, with concerns about privacy and the spread of misinformation. Social media platforms have come under fire for allowing the spread of fake news and hate speech, and for failing to protect user data.

Despite these concerns, social media continues to be a major force in society and business. The rise of social media has changed the way we communicate, connect, and do business, and it will likely continue to evolve and shape our lives in the years to come.

The emergence of social media has changed the way we communicate and interact with each other. In recent years, businesses have also been leveraging social media platforms to reach out to customers and promote their products and services. However, the vast amount of data generated by social media can be challenging to manage, leading to the rise of cloud computing technology.

Cloud computing is a model of delivering computing services over the internet. It allows users to access shared computing resources, such as servers, storage, and applications, on demand. The history of cloud computing can be traced back to the 1960s, when the concept of time-sharing emerged. This allowed multiple users to access a single computer at the same time, thereby making more efficient use of computing resources.

The development of cloud computing technology can be attributed to the increased availability of high-speed internet connections and advances in virtualization technology. Virtualization allows multiple virtual machines to run on a single physical machine, making it possible to share computing resources more efficiently.

The impact of cloud computing on business and society has been profound. It has enabled businesses to scale their operations quickly and efficiently, without the need for significant upfront investment in hardware and infrastructure. Cloud computing has also made it easier for businesses to collaborate and share information, regardless of their physical location.

One of the potential benefits of cloud computing is cost savings. Since businesses only pay for the computing resources they use, they can avoid the high upfront costs associated with building and maintaining their own IT infrastructure. Cloud computing also allows businesses to scale up or down quickly, depending on their changing needs, without incurring significant additional costs.

Scalability is another significant benefit of cloud computing. Businesses can quickly add or remove computing resources as needed, without having to worry about the limitations of their physical infrastructure. This allows businesses to respond quickly to changing market conditions and customer needs.

Cloud computing also provides greater accessibility to computing resources. Users can access the cloud from anywhere in the world, as long as they have an internet connection. This allows businesses to offer their services to customers around the world, without having to set up physical infrastructure in each location.

The development of cloud computing technology has had a significant impact on business and society. It has enabled businesses to scale their operations quickly and efficiently, while also providing greater accessibility and cost savings. However, the challenges of data security and privacy must be addressed to ensure the continued growth and adoption of cloud computing.

As businesses and organizations increasingly adopted cloud computing, the amount of data generated skyrocketed. This created a need for tools and technologies to help manage and make sense of the vast amounts of data being generated. Enter big data.

The term "big data" refers to data sets that are too large and complex to be processed by traditional data processing applications. Big data can come from a variety of sources, including social media, online transactions, and even sensors on connected devices.

The emergence of big data presented both challenges and opportunities. On the one hand, businesses now had access to a wealth of information that could be used to improve operations, enhance customer experiences, and drive growth. On the other hand, managing and analyzing large data sets was a daunting task that required new tools and approaches.

The rise of big data led to the development of analytics, which involves using data to derive insights and make decisions. Analytics can take many forms, including descriptive analytics, which looks at historical data to understand what happened in the past, and predictive analytics, which uses machine learning algorithms to forecast future outcomes based on historical data.

The potential impact of big data and analytics on business and society is significant. By leveraging data insights, businesses can identify areas for improvement, optimize operations, and even create new products and services. For example, retailers can use customer data to personalize marketing campaigns and improve the customer experience, while healthcare providers can use patient data to identify trends and develop new treatments.

The rise of big data and analytics has transformed the way businesses operate and has the potential to revolutionize society as a whole. By leveraging data insights, businesses can drive innovation, improve operations, and enhance customer experiences. However, this requires overcoming significant challenges around data quality, privacy, and bias. As the amount of data being generated continues to grow, it is essential that businesses and organizations invest in the tools and talent needed to manage and analyze it effectively.

Now let's quickly pivot the conversation to the emergence of digital media, which has been a significant part of the transformation of the media and entertainment industry. With the advent of the internet and new technologies, traditional media, such as newspapers and magazines, have seen a decline in their influence and revenues. The internet has made it possible for anyone to access news and information instantly, and social media platforms have become the go-to source for breaking news and updates.

However, with the rise of digital media, new opportunities have also emerged. Streaming services, such as Netflix and Spotify, have become a significant player in the entertainment industry, offering consumers personalized content and an on-demand experience. Netflix, in particular, has disrupted the traditional television industry, with its subscription-based model and original content productions.

The impact of digital media on the entertainment industry has been significant. Streaming services have changed the way people consume media, with binge-watching becoming a common practice. The availability of personalized content has also led to a rise in niche programming, with streaming services catering to specific audiences based on their interests and preferences.

The growth of digital media has also had an impact on the advertising industry. Personalized advertising based on user data has become the norm, with companies using data analytics to target specific audiences with tailored messaging.

Overall, the rise of digital media has transformed the media and entertainment industry, offering new opportunities and challenges. While traditional media may have seen a decline in their influence and revenues, streaming services have emerged as a new player in the industry, providing consumers with personalized content and on-demand experiences. The impact of digital media on the advertising industry has also been significant, with personalized advertising becoming the norm, but also raising concerns about privacy and data protection. As technology continues to evolve, it will be interesting to see how the media and entertainment industry adapts and transforms in response.

As we move into the present day, the history of digital transformation brings us to the exciting and rapidly-evolving field of artificial intelligence and machine learning. AI is the ability of machines to simulate human intelligence and perform tasks that typically require human cognitive abilities, such as visual perception, speech recognition, decision-making, and natural language processing. Machine learning is a subset of AI that involves training computers to learn from data and improve over time, without being explicitly programmed.

The idea of AI dates back to the mid-20th century, when pioneers such as Alan Turing and John McCarthy began to explore the concept of machines that could think and reason like humans. However, it wasn't until the digital era that AI really began to take off. In the 1980s and 1990s, there was a surge of interest in expert systems, which were computer programs designed to mimic the decision-making abilities of human experts in a particular field. Expert systems paved the way for more advanced forms of AI, such as neural networks and deep learning, which use layers of artificial neurons to analyze complex data and recognize patterns.

Today, AI and machine learning are being used in a wide range of applications, from self-driving cars and facial recognition technology to personalized recommendation systems and predictive analytics. In healthcare, AI is being used to improve diagnoses and treatment plans, while in manufacturing, it's helping to optimize supply chain management and quality control. In finance, AI is being used to detect fraud and make investment decisions, and in customer service, it's being used to provide personalized support and improve user experiences.

However, the rapid advancement of AI and machine learning also raises important ethical and societal concerns. One of the biggest challenges is ensuring that these technologies are used in a responsible and transparent way, without perpetuating biases or causing harm. There are also concerns about the impact of AI on the workforce, with some experts predicting that automation and AI could displace millions of jobs in the coming years.

Now, digital transformation has had a significant impact on the education sector as well, with the rise of e-learning and online education platforms. The history of e-learning can be traced back to the 1980s when computer-based training programs were first introduced. These programs were typically delivered on floppy disks or CD-ROMs and provided a self-paced learning experience.

However, it wasn't until the rise of the internet that online education truly took off. In the late 1990s and early 2000s, universities and other educational institutions began to offer online courses and degree programs. The first online course, offered by the University of Phoenix, launched in 1989.

The emergence of Massive Open Online Courses (MOOCs) in the early 2010s further revolutionized the online education landscape. These courses, offered by universities and other educational organizations, were often free and open to anyone with an internet connection. This democratized access to education and made it possible for people around the world to learn from top educators and institutions.

Today, online education platforms like Coursera and edX offer a wide range of courses and degree programs from top universities and institutions around the world. These platforms have enabled millions of learners to access high-quality education and gain new skills and knowledge.

One of the main benefits of online education is its accessibility. Online courses and programs can be accessed from anywhere in the world, at any time. This makes education more accessible to people who may not have the opportunity to attend traditional brick-and-mortar schools, such as those in remote or underserved areas.

Online education also has the potential to be more affordable than traditional education. Online courses and programs often have lower tuition fees and don't require the additional costs associated with attending a physical campus, such as housing and transportation.

However, online education also presents some challenges. Quality control can be a concern, as there is often no way to ensure that learners are actually completing the coursework or assessments themselves. There is also the issue of ensuring that online courses and programs are of the same quality as traditional programs, and that learners are receiving the same level of instruction and support.

As we can see, digital technology has also changed the way we learn and teach, with the rise of virtual and augmented reality for immersive learning experiences. These technologies allow learners to engage with content in a more interactive and engaging way, which can improve learning outcomes and retention.

In a nutshell, digital transformation has had a profound impact on education, with the rise of e-learning and online education platforms. These platforms have enabled greater accessibility to education and made it possible for people around the world to learn from top educators and institutions. With the continued advancement of digital technology, it will be interesting to see how the education landscape continues to evolve in the coming years.

Speaking of evolution, the Internet of Things (IoT) is a term that refers to the interconnection of various devices and objects through the internet, allowing them to communicate with each other and share data. The concept of IoT can be traced back to the late 1990s, but it was not until the 2010s that it started to gain traction.

In the early days of IoT, it was primarily used in industrial and commercial settings, with sensors and other devices being used to monitor and control processes in factories, warehouses, and other facilities. These early IoT systems were often proprietary and used their own communication protocols, which made it difficult to connect different devices and systems.

However, with the growth of the internet and wireless networks, IoT has become more accessible and widespread. Today, IoT devices can be found in homes, cars, and even wearables like fitness trackers and smartwatches.

One of the key benefits of IoT is its ability to collect and analyze data in real-time. For example, in a smart home, sensors can be used to monitor temperature, humidity, and other environmental factors, allowing the system to automatically adjust heating, cooling, and lighting to optimize energy usage and improve comfort.

Similarly, in a smart city, IoT sensors can be used to monitor traffic flow, air quality, and other factors to improve transportation, reduce congestion, and enhance public safety.

However, with the growth of IoT comes new challenges, particularly in terms of security and privacy. Because IoT devices are often connected to the internet and communicate with other devices and systems, they are vulnerable to cyber-attacks and data breaches.

In addition, because IoT devices often collect and transmit sensitive data, such as personal health information or financial data, there are concerns about the potential misuse of this data and the need to ensure that proper security measures are in place.

Despite these challenges, the potential benefits of IoT are significant, and it is likely that we will continue to see rapid growth and adoption of this technology in the coming years. As more devices become connected and more data is collected and analyzed, we can expect to see new innovations and opportunities for businesses, governments, and individuals alike.

Last, but definitely not least, quantum computing is a relatively new and rapidly evolving field that promises to revolutionize industries across the board. It has the potential to solve complex problems and perform calculations that traditional computers would take millions of years to complete. Quantum computing takes advantage of the unique properties of quantum mechanics to process information and solve problems in new and innovative ways.

The history of quantum computing can be traced back to the early 1980s when physicist Richard Feynman proposed the idea of a quantum computer. However, it wasn't until the late 1990s that the first experimental quantum computers were built. These early machines were rudimentary and could only perform simple calculations, but they demonstrated the potential of quantum computing technology.

In the years that followed, quantum computing technology advanced rapidly, and researchers began exploring its potential applications. One of the most promising applications of quantum computing is in the field of drug discovery. With quantum computers, scientists can simulate the behavior of molecules and design new drugs that are more effective and have fewer side effects. This could have a significant impact on the pharmaceutical industry and help to address some of the world's most pressing health challenges.

Quantum computing also has the potential to revolutionize the field of finance. With quantum computers, analysts can perform complex financial modeling and risk analysis in real-time, enabling them to make better investment decisions and reduce financial risk. This could have a significant impact on the global economy and the financial industry.

The challenges of developing and scaling quantum computing technology are significant. One of the biggest challenges is developing qubits, the basic building blocks of quantum computers. Qubits are highly sensitive and easily disturbed by their environment, making them difficult to manipulate and control. Researchers are working to develop new materials and technologies that can overcome these challenges and enable the development of large-scale quantum computers.

Another challenge is the need for new software and programming languages that can take advantage of quantum computing technology. Traditional programming languages are not suitable for quantum computing, and researchers are developing new programming languages and algorithms that can work with qubits and take advantage of their unique properties.

Despite these challenges, the potential benefits of quantum computing are significant, and researchers and businesses around the world are investing heavily in its development. The race to build the first practical quantum computer is on, and the winner could change the course of history.

The history of quantum computing is relatively short, but its potential impact on industries such as healthcare, finance, and information technology are enormous. The development and scaling of quantum computing technology present significant challenges, but researchers and businesses are working tirelessly to overcome these challenges and unlock the full potential of this groundbreaking technology. The future of quantum computing is exciting and full of possibilities, and its impact on society and the world as we know it could be truly transformative.

In conclusion, the history of digital transformation has seen many significant technological advancements that have transformed society and industries. The early days of computing saw the invention of the first computers and the development of programming languages. The rise of the internet and the creation of the World Wide Web led to the growth of e-commerce and online communication. The introduction of smartphones and mobile apps revolutionized the way we communicate and do business. Social media platforms have changed the way we connect and interact with each other. Cloud computing has enabled organizations to scale and save costs, but also raised concerns around data security and privacy. Big data and analytics have transformed the way businesses make decisions. The entertainment industry has also undergone significant change with the rise of digital media and streaming services. Artificial intelligence and machine learning have the potential to revolutionize various industries. Online education platforms have made education more accessible yet also present challenges around quality control. The Internet of Things (IoT) has the potential to connect devices and data in real-time. Lastly, quantum computing has emerged as a potential game-changer.

What we will visit in the next chapter is how in today's modern society, digital transformation continues to impact our lives and change the way we live, work, and interact with each other. The integration of digital technology is crucial for businesses to remain competitive and relevant, while also presenting opportunities for innovation and growth. As we move forward, it is important to address the challenges that arise with digital transformation, including the ethical and responsible use of technology, ensuring data privacy and security, and addressing concerns around automation and job displacement. By doing so, we can harness the potential of digital transformation to create a better future for all.

CHAPTER THREE: DIGITALIZATION IN MODERN SOCIETY

The world is undergoing a rapid transformation, fueled by advancements in technology that have revolutionized the way we live, work, and communicate. The term "digital transformation" has become a key term, describing the process by which organizations and individuals are adapting to this new reality.

As mentioned in the introduction, digital transformation is not just about adopting new technologies or updating old ones. It is about fundamentally changing the way we do things, from the way we shop, to the way we learn, to the way we govern ourselves. This transformation is driven by a number of factors, including the rise of mobile computing, the proliferation of data, and the advent of artificial intelligence and machine learning.

At the heart of digital transformation is the idea that we can use technology to improve our lives in countless ways. We can make better decisions, connect with others more easily, and create new products and services that were once impossible. We can also become more efficient and effective in our work, reducing costs and improving the quality of our output.

But digital transformation is not without its challenges. As we become more dependent on technology, we also become more vulnerable to its failures. Cybersecurity threats and data breaches are on the rise, and the risks associated with these events can be catastrophic.

Moreover, digital transformation has profound implications for society as a whole. It can exacerbate inequality, as those who lack access to technology are left behind. It can also have unintended consequences, such as the loss of jobs and the erosion of privacy.

In this book chapter, we will explore the many facets of digital transformation, and its impact on modern society. We will examine how technology is transforming the way we live, work, and communicate, and explore the opportunities and challenges this presents. We will also consider how digital transformation is changing the way we think about governance, democracy, and human rights.

As we have explored in the previous sections, digital transformation has had a profound impact on many aspects of modern society, from the way we live our lives to the way we work and govern ourselves. One area where this impact is particularly noticeable is in the way we communicate with each other.

Communication technology has been evolving for centuries, from the first postal service in ancient Rome to the telegraph and telephone in the 19th century. However, the rise of digital technology has taken this evolution to a whole new level. Today, we have access to an unprecedented range of communication tools, from email and instant messaging to video conferencing and social media.

This has had a profound impact on the way we interact with each other. In the past, communication was largely limited to those who were physically present or within shouting distance. Today, we can communicate with people on the other side of the world instantaneously, using a wide range of tools and platforms.

Social media, in particular, has had a profound impact on communication. Platforms like Facebook, Twitter, and Instagram have made it easier than ever to connect with friends, family, and colleagues, and to share our thoughts and experiences with a wider audience. Social media has also given rise to new forms of communication, such as memes and hashtags, that are unique to the digital age.

However, social media has also had its downsides. The rise of fake news and online misinformation has made it more difficult to separate fact from fiction, and has contributed to the polarization of society. Social media has also been linked to mental health issues, such as depression and anxiety, particularly among young people.

The use of digital technology in personal relationships can also have drawbacks. For one, it can lead to a sense of disconnection and depersonalization in personal communication. When communication is limited to text messages or social media posts, it can be difficult to convey emotions and express oneself fully. This can lead to misunderstandings and a lack of depth in personal relationships.

Additionally, the constant use of digital technology can also lead to a sense of overload and burnout, which can negatively impact personal relationships. When individuals are constantly connected to their devices and are bombarded with notifications and messages, it can be difficult to fully engage in personal interactions and be present in the moment.

Furthermore, the use of digital technology in personal relationships can also lead to issues of trust and privacy. With the ability to track and monitor communication through social media platforms and messaging apps, individuals may feel their privacy is being compromised. This can lead to feelings of anxiety and distrust in personal relationships, which can negatively impact the overall quality of the relationship.

Despite these challenges, there is no doubt that digital technology has revolutionized the way we communicate. It has made it easier than ever to connect with others, and has given rise to new forms of communication that were once impossible. As we continue to adapt to this new reality, it is important that we remain vigilant to the risks and challenges that digital communication poses, while also embracing its potential for positive change.

Real-world examples of digital technology's impact on communication are plentiful. For instance, during the COVID-19 pandemic, digital communication tools such as Zoom and Microsoft Teams became essential for remote work and virtual meetings. Similarly, social media played a crucial role in enabling people to stay connected with loved ones during lockdowns and travel restrictions. The rise of citizen journalism and the use of social media to document events like protests and natural disasters has also had a significant impact on the way we consume news and information.

In addition to social media, there are a wide range of modern-day communication technologies that have emerged in recent years as a result of digital transformation. These technologies have enabled us to communicate more easily, more frequently, and with greater speed and efficiency than ever before.

One example is email, which has been around for several decades but continues to be a vital tool for communication. Email enables us to send messages and attachments to anyone with an email address, anywhere in the world, instantly. This has transformed the way we communicate in both our personal and professional lives, making it easier to stay in touch with colleagues, friends, and family members who are far away.

Another example is instant messaging, which has become increasingly popular in recent years. Apps like WhatsApp, Facebook Messenger, and WeChat enable us to send text messages, voice messages, and even make phone and video calls, all from a single platform. This has made communication more convenient and efficient, particularly for people who are on the go or who need to communicate with people in different time zones.

Video conferencing is another technology that has become increasingly important in recent years. Platforms like Zoom and Microsoft Teams have become essential tools for remote work and virtual meetings, particularly during the COVID-19 pandemic. Video conferencing enables us to communicate face-to-face with people anywhere in the world, without the need for expensive travel or complicated technical setups.

Collaboration tools are also becoming more significant in modern communication. Tools like Google Docs and Dropbox enable people to work together on projects in real-time, even if they are in different parts of the world. This has transformed the way we work, making it easier to collaborate with colleagues and partners across different time zones and locations.

Finally, voice assistants like Amazon Alexa, Google Home, and Apple HomePod have become increasingly popular in recent years. These devices enable us to control our homes, access information, and communicate with other devices using voice commands. This hands-free communication technology has become especially valuable for people with disabilities or mobility issues, as it allows them to control their environment and access information without relying on physical devices. Voice assistants have also become an important tool for businesses, enabling them to provide customers with a more personalized experience and streamline their operations. For example, restaurants can use voice assistants to take orders and provide recommendations, while retailers can use them to assist customers with finding products and making purchases. As voice assistants continue to improve and become more integrated into our daily lives, it is likely that they will play an even greater role in shaping the future of communication technology.

The evolution of communication technology and the rise of digital transformation has transformed the way we communicate with each other. While there are certainly risks and challenges associated with this new reality, there is also great potential for positive change. As we continue to adapt to this new reality, it is important that we remain mindful of the impact of digital communication on our lives and society as a whole.

Another area where digital transformation has had a significant impact is education. Digital technology has transformed the way we teach and learn, making education more accessible and convenient than ever before.

One of the most significant changes in education has been the growth of online learning platforms. Platforms like Coursera, Udemy, and edX have made it possible for anyone with an internet connection to access high-quality educational content from some of the world's leading universities and institutions. This has made education more accessible to people in remote or underserved areas, and has opened up new opportunities for lifelong learning and professional development.

In addition to online learning platforms, digital technology has also had a profound impact on the way we teach and learn. Interactive whiteboards, tablets, and other digital tools have made it possible for teachers to create more engaging and interactive learning experiences, while online resources and digital textbooks have made it easier for students to access and organize information.

Digital technology has also enabled new forms of collaborative learning. Online discussion forums, video conferencing, and collaborative editing tools have made it easier for students to work together on group projects and assignments, regardless of their physical location.

However, there are also challenges associated with the use of digital technology in education. One of the biggest challenges is the need to ensure that all students have access to the technology and resources needed to participate fully in digital learning experiences. This can be particularly challenging in low-income or underserved communities, where access to technology and internet connectivity may be limited.

Another challenge is the need to ensure that digital learning experiences are engaging and effective. While digital tools can enhance learning in many ways, they can also be a source of distraction and can sometimes detract from the overall learning experience. It is important for educators to strike the right balance between using digital technology to enhance learning, while also ensuring that students remain engaged and focused on the learning objectives.

In addition to online learning platforms and digital textbooks, there are a variety of other digital tools that are used in education to enhance teaching and learning. These tools have transformed the way that educators approach curriculum design, lesson planning, and classroom management.

One of the most popular digital tools in education is the interactive whiteboard. Interactive whiteboards allow teachers to create dynamic and engaging lessons that incorporate multimedia elements like videos, animations, and interactive exercises. Teachers can use these tools to annotate presentations, highlight key concepts, and incorporate student feedback in real-time.

Tablets and other mobile devices have also become increasingly popular in the classroom. These devices allow students to access course materials, collaborate on group projects, and complete assignments from anywhere, at any time. Many schools have implemented "bring your own device" (BYOD) policies, allowing students to use their own smartphones, tablets, or laptops in the classroom.

Digital simulations and virtual reality (VR) experiences have also become popular tools for teaching complex concepts in a hands-on, immersive way. For example, a biology teacher might use a digital simulation to help students visualize the workings of the human body at a molecular level, or a history teacher might use a VR experience to transport students to a historic event or location.

Other digital tools used in education include online discussion forums, video conferencing platforms, and collaborative editing tools. These tools allow students to work together on group projects and assignments, regardless of their physical location. They also facilitate communication and collaboration between teachers and students, allowing educators to provide more personalized feedback and support.

Digital assessment tools are another important aspect of modern education. These tools allow teachers to administer quizzes and tests electronically, reducing the time and resources required for grading and analysis. Digital assessment tools can also provide more detailed feedback on student performance, helping teachers identify areas where individual students may need extra support or intervention.

Overall, digital tools have transformed the way that educators approach teaching and learning. By incorporating multimedia elements, facilitating collaboration, and providing personalized feedback, these tools have made education more engaging, accessible, and effective for students of all ages and abilities. As technology continues to evolve, it is likely that we will see even more innovative digital tools emerge in the education space, further transforming the way we approach teaching and learning.

The impact of digital technology on education has been significant. From online learning platforms to interactive whiteboards, digital technology has transformed the way we teach and learn. While there are certainly challenges associated with the use of digital technology in education, there is also great potential for positive change. As we continue to adapt to this new reality, it is important that we work to ensure that all students have access to the technology and resources they need to participate fully in digital learning experiences, while also ensuring that these experiences remain engaging, effective, and focused on learning objectives.

Another field where digital technology has made significant strides is healthcare. From telemedicine to electronic health records, digital technology has transformed the way we approach healthcare delivery.

The use of technology in healthcare has enabled healthcare providers to deliver care more efficiently and effectively. Telemedicine has made it possible for doctors and patients to connect remotely, allowing patients to receive care from the comfort of their own homes. Electronic health records have made it easier for healthcare providers to access and share patient information, improving coordination of care and reducing the likelihood of medical errors.

Digital technology has also had a significant impact on patient care and outcomes. Wearable devices and other digital health tools have made it possible for patients to monitor their own health and track their progress towards wellness goals. Remote monitoring has also made it possible for healthcare providers to monitor patients with chronic conditions and intervene early to prevent complications.

In addition to improving patient care, digital technology has also helped to reduce healthcare costs. The use of electronic health records has streamlined administrative tasks, reduced paperwork and improved efficiency. Remote monitoring and telemedicine have also reduced the need for in-person visits, lowering healthcare costs for both patients and providers.

Real-world examples of the impact of digital technology on healthcare abound. For example, wearable devices like Fitbits and Apple Watches have become increasingly popular among consumers, allowing people to track their own health and wellness. Telemedicine has also become more common, particularly during the COVID-19 pandemic, allowing doctors to connect with patients remotely and reducing the risk of exposure to the virus.

Digital technology in healthcare has been significant. From telemedicine to wearable devices, digital technology has transformed the way we approach healthcare delivery and patient care. While there are certainly challenges associated with the use of digital technology in healthcare, there is also great potential for positive change. As we continue to innovate and adapt to this new reality, it is important that we work to ensure that all patients have access to the technology and resources they need to receive high-quality care, while also maintaining the human touch that is so essential to the practice of medicine.

Now, the entertainment industry has been one of the most drastically affected by the rise of digital technology. Streaming services such as Netflix, Hulu, and Amazon Prime have completely revolutionized the way we watch movies and TV shows. Now, instead of being tied to a cable subscription, we can access an extensive library of content anytime and anywhere with an internet connection.

And it's not just the variety of content that has changed, but the way we pay for it. The subscription-based pricing model has made it more accessible and affordable for people to access their favorite shows and movies. Plus, the introduction of original content production by these platforms has given us even more options to choose from.

The music industry has also undergone a significant transformation thanks to digital technology. Gone are the days of purchasing individual songs or albums. Nowadays, we can access millions of songs with just a monthly subscription fee on platforms like Spotify and Apple Music. This has made it easier for artists to distribute their music and reach a global audience without the need for physical distribution networks.

Even the gaming industry has seen significant growth with the advent of digital technology. Mobile gaming has become a billion-dollar industry, and online multiplayer games have given rise to esports, where professional gamers can compete for huge sums of money.

Examples of these changes are all around us. Disney+, for instance, shook up the streaming industry with their impressive library of content and original programming. And the music industry is barely recognizable from what it used to be, with streaming platforms now dominating the way we listen to and distribute music.

Of course, there are challenges associated with the growth of digital entertainment, such as the need to address piracy and copyright issues. However, there is great potential for positive change in the industry, and we must work together to ensure that everyone involved can benefit from these opportunities.

Digital technology has not only changed the way we access entertainment content but also how it is produced and distributed. From movie production to music recording, technology has revolutionized the entertainment industry in countless ways.

One of the most significant technological advancements in the entertainment industry is computer-generated imagery (CGI). This technology has allowed filmmakers to create visually stunning effects and even entire worlds that would be impossible to film in real life. Movies like Avatar and The Lord of the Rings trilogy are great examples of how CGI has transformed the film industry.

Virtual reality (VR) is another technology that is rapidly changing the way we experience entertainment. VR headsets and gaming systems have given us immersive experiences in which we can interact with digital environments in a more realistic way. This technology is particularly popular in the gaming industry, but its potential applications go far beyond gaming. For instance, it could be used to create virtual tours of museums or historical sites, allowing people to explore these places in ways that were not possible before.

Augmented reality (AR) is another technology that is transforming the entertainment industry. AR technology overlays digital information on the real world, providing a unique and interactive experience for users. For instance, the popular game Pokemon Go uses AR to allow players to catch virtual creatures in the real world.

Live streaming is another example of digital technology transforming the entertainment industry. Musicians can now live stream their concerts and reach a global audience, while gamers can live stream their gameplay on platforms like Twitch and connect with fans in real-time.

Lastly, social media has transformed the way we interact with entertainment content. Fans can now engage with their favorite artists and performers on social media platforms like Twitter, Instagram, and TikTok. This has given fans a more direct connection with the entertainment industry and has allowed for new forms of fan engagement.

In summary, digital technology has transformed the entertainment industry in countless ways. From CGI to virtual reality, these technologies have opened up new possibilities for content creation and distribution. As technology continues to evolve, it's exciting to imagine what new innovations will shape the entertainment industry in the years to come.

Let's pivot briefly...transportation is an important aspect of our daily lives, and digital technology has had a significant impact on it. With the advent of ride-sharing services like Uber and Lyft, getting around has become more accessible and convenient for many people. No longer do we have to wait on street corners for a taxi, or own a car to get where we need to go. Ride-sharing services have made it possible for us to travel without the burden of car ownership.

The growth of ride-sharing services has also had a positive impact on the environment. By reducing the number of cars on the road, ride-sharing services have helped to reduce traffic congestion and improve air quality in many cities. And with the introduction of electric vehicles, the environmental benefits of ride-sharing services are even more significant.

But the impact of digital technology on transportation doesn't stop there. The development of autonomous vehicles has the potential to revolutionize the way we think about transportation. Imagine being able to sit back and relax on your morning commute, while your car drives itself to work. This could become a reality in the not-too-distant future, as companies like Tesla and Google continue to invest in autonomous vehicle technology.

The history of electric and autonomous vehicles is a fascinating one, spanning over a century of innovation and development. The idea of using electric power as a means of propulsion for vehicles dates back to the early days of the automobile industry, and the development of autonomous vehicles has been a dream of scientists and engineers for even longer.

The first electric car was built in 1837 by Scottish inventor Robert Anderson. However, it was not until the late 1800s that electric vehicles (EVs) began to gain popularity as a viable alternative to gasoline-powered cars. At the time, EVs were particularly popular among women, who preferred the cleaner and quieter operation of electric cars over the loud and smelly gasoline cars. In fact, in 1900, electric vehicles accounted for about one-third of all cars on the road in the United States.

One of the pioneers of the electric car industry was Thomas Davenport, an American inventor who built a small electric vehicle in 1835. Another early innovator was William Morrison, who built the first successful electric car in the United States in 1891. However, it was the development of the lead-acid battery in the early 1900s that really propelled the electric car industry forward. With these batteries, electric cars could travel longer distances and at higher speeds than ever before.

Despite their early success, the electric car industry was dealt a major blow with the introduction of Henry Ford's Model T in 1908. The Model T was cheaper and more practical than electric cars, and quickly became the dominant form of transportation in the United States. As a result, the electric car industry went into decline, and it was not until the 1990s that electric cars began to make a comeback.

In the 1990s, California passed the Zero Emission Vehicle (ZEV) mandate, which required automakers to produce a certain percentage of zero-emission vehicles, including electric cars. This led to the development of the General Motors EV1, the first mass-produced electric car, which was introduced in 1996. However, the EV1 was ultimately discontinued in 2003 due to a lack of consumer demand.

It was not until the mid-2010s that electric cars began to gain real traction in the market. The introduction of the Tesla Model S in 2012 helped to change the perception of electric cars from being slow and impractical to being stylish and high-performing. Today, electric cars are becoming increasingly popular, with many major automakers introducing their own electric models.

The idea of a self-driving car dates back to at least the 1920s, when futurists and science fiction writers began to imagine a world in which cars could drive themselves. However, it was not until the 1980s and 1990s that serious research into autonomous vehicles began.

One of the early pioneers of autonomous vehicle research was Ernst Dickmanns, a German scientist who developed a self-driving car called the VaMP in the 1980s. The VaMP used cameras and sensors to navigate the roads, and was able to drive up to 60 miles per hour on the highway. Another early innovator was the Defense Advanced Research Projects Agency (DARPA), which sponsored a series of autonomous vehicle competitions in the early 2000s. These competitions helped to push the development of autonomous vehicles forward, and paved the way for more advanced research and development.

In recent years, companies like Google and Tesla have been at the forefront of autonomous vehicle research and development. Google's self-driving car project, Waymo, has been testing autonomous vehicles on public roads since 2009

The potential impact of autonomous vehicles on transportation and society is significant. It could reduce the number of accidents caused by human error, increase efficiency and reduce traffic congestion, and provide a more convenient and accessible way to travel for people who cannot drive.

Another technological innovation that is transforming transportation is the growth of electric scooters and bikes. Companies like Lime and Bird have introduced electric scooters that can be rented on demand, providing a convenient and eco-friendly way to travel short distances in urban areas. Similarly, electric bikes have become increasingly popular, with bike-sharing programs allowing people to rent bikes for short periods of time.

As technology continues to evolve, we can expect to see even more changes in the way we think about transportation. From hyperloops to flying cars, there are many exciting and ambitious projects in the works that could change the way we travel. However, it is important to ensure that these technologies are developed in a responsible and sustainable way, with consideration for their impact on society and the environment.

The impact of digital technology is not limited to the private sector. In recent years, governments around the world have increasingly recognized the potential of digital technology to improve the efficiency and transparency of public services. The adoption of e-government services has been driven by the desire to provide citizens with convenient and accessible services while also reducing the costs of service delivery.

E-government services refer to the delivery of government services and information to citizens through electronic means. These services may include online tax filing, online voting, online payment of bills, and online application for government programs. The use of digital technology in government services has been growing rapidly in recent years, with many governments adopting a "digital-first" approach to service delivery.

One example of successful e-government implementation is Estonia, which has been recognized as a global leader in e-government services. The country has a comprehensive national electronic ID system that allows citizens to access a wide range of government services online, from voting to healthcare. The system has significantly improved the efficiency of public services and reduced bureaucracy. In addition, it has increased transparency by making government processes more accessible and visible to citizens.

Another example is Singapore, which has implemented a range of e-government services to improve citizen engagement and service delivery. The government's "Smart Nation" initiative aims to leverage digital technology to transform the city-state into a smart nation, where technology is used to enhance quality of life, improve productivity, and create economic opportunities.

The initiative was launched in 2014 with the aim of enhancing the quality of life of Singaporeans, improving productivity, and creating economic opportunities. The Smart Nation Initiative covers a broad range of areas, including transport, healthcare, education, and public services.

The initiative is built on three key pillars: digital infrastructure, digital economy, and digital government. The digital infrastructure pillar aims to provide a strong foundation for digital technology by investing in high-speed connectivity, cloud computing, and data analytics. The digital economy pillar aims to create opportunities for businesses and entrepreneurs by promoting innovation, skills development, and internationalization. The digital government pillar aims to transform the way government services are delivered by making them more citizen-centric, efficient, and responsive.

One key initiative under the Smart Nation Initiative is the National Digital Identity (NDI) program, which aims to provide citizens and businesses with a secure and convenient way to access digital services. The NDI program enables citizens to use a single digital identity across multiple government services, reducing the need to remember multiple usernames and passwords. The program also uses advanced security features such as biometrics to ensure the identity of users is protected.

Another key initiative is the Smart Nation Sensor Platform (SNSP), which aims to leverage the Internet of Things (IoT) to improve the efficiency and sustainability of urban infrastructure. The SNSP involves the deployment of sensors across the city-state to collect data on a range of factors such as traffic flow, air quality, and water levels. This data is then used to inform decision-making and improve the efficiency of services such as transport and waste management.

The Smart Nation Initiative has also led to the development of innovative digital services such as the Virtual Singapore platform, which provides a 3D digital twin of the city-state that can be used for urban planning and development. The platform allows users to simulate the impact of proposed changes to the cityscape and infrastructure, helping to improve decision-making and enhance the sustainability of the built environment.

The Smart Nation Initiative has already had a significant impact on Singaporean society. For example, the adoption of e-government services has made it easier for citizens to access government services and information, reducing the need for physical visits to government offices. The NDI program has also made it easier for businesses to interact with the government by streamlining the process of accessing permits and licenses.

The Smart Nation Initiative has also led to the creation of new jobs and economic opportunities in the digital sector. The government has invested heavily in skills development and education to ensure that Singaporeans are equipped with the necessary digital skills to thrive in the digital economy.

Overall, the Smart Nation Initiative represents a bold and ambitious vision for the future of Singapore. By leveraging digital technology to improve the quality of life of citizens, enhance productivity, and create economic opportunities, the initiative has the potential to transform Singapore into a leading smart city and an example for other countries to follow.

As we can see, the adoption of e-government services has the potential to improve government efficiency and reduce costs. By providing citizens with access to government services online, governments can reduce the need for physical infrastructure and staff, which can lead to significant cost savings. In addition, the use of digital technology can streamline government processes and reduce bureaucracy, which can improve efficiency and responsiveness.

Furthermore, e-government services can also improve the accessibility and convenience of government services for citizens. By allowing citizens to access services online, governments can provide services 24/7, regardless of location. This can be especially beneficial for citizens in remote or rural areas who may have limited access to physical government offices.

However, there are also concerns about the potential drawbacks of e-government services. One major concern is the potential for privacy breaches and data security issues. As citizens provide personal information online, there is a risk that this information may be accessed by unauthorized individuals, leading to identity theft and other security issues.

Another concern is the potential for exclusion of citizens who may not have access to digital technology or who may lack the necessary digital literacy skills to use e-government services effectively. This could create a "digital divide" between those who have access to digital technology and those who do not, exacerbating existing social and economic inequalities.

Despite these concerns, the potential benefits of e-government services are significant. By improving government efficiency and transparency, e-government services can lead to greater trust and engagement between citizens and their governments. In addition, by improving the accessibility and convenience of government services, e-government services can help to increase citizen participation in democracy and enhance the quality of life for all citizens.

The adoption of e-government services represents a significant opportunity for governments around the world to improve the efficiency and transparency of public services. By leveraging digital technology to provide citizens with accessible and convenient services, governments can reduce costs, streamline processes, and enhance citizen engagement. However, in order to fully realize the potential of e-government services, governments must also address concerns around privacy and digital inclusion, and work to ensure that all citizens have equal access to the benefits of digital technology.

To piggy back off the topic of public services, its only right to touch upon technology and social justice. The impact of digital technology on social justice has been a topic of much discussion in recent years. On one hand, digital technology has the potential to reduce inequality and increase access to resources for marginalized communities. On the other hand, there are concerns that relying too heavily on digital technology to address social justice issues may create new forms of inequality and exacerbate existing ones.

One potential benefit of digital technology in promoting social justice is the increased access to resources it can provide. For example, online learning platforms can provide educational opportunities to individuals who may not have had access to traditional education due to financial or geographical constraints. This can help to level the playing field and provide opportunities for upward mobility.

Digital technology can also facilitate communication and collaboration among social justice advocates and organizations. Social media platforms, for example, have been used to raise awareness of social justice issues and mobilize individuals to take action. The #MeToo movement, which began as a hashtag on social media, is one such example. The movement brought attention to the prevalence of sexual harassment and assault and led to changes in policies and cultural attitudes towards these issues.

However, there are potential drawbacks to relying too heavily on digital technology to address social justice issues. One concern is the digital divide, which refers to the unequal distribution of access to digital technology. Individuals and communities who lack access to technology may be left behind and unable to take advantage of the opportunities provided by digital technology. This can further exacerbate existing inequalities.

Another concern is the potential for digital technology to perpetuate existing biases and inequalities. For example, algorithms used in hiring or lending decisions may inadvertently discriminate against marginalized communities if the algorithms are not designed with diversity and inclusivity in mind. This highlights the importance of ensuring that digital technology is designed and implemented in a way that is equitable and inclusive.

Furthermore, there is a risk that reliance on digital technology may lead to a depersonalization of social justice issues. The ease of sharing and consuming information through digital technology may lead to a superficial understanding of complex social justice issues. This can result in a lack of nuance and understanding of the complexities of social justice issues, leading to oversimplification and potentially harmful solutions.

Digital technology has the potential to be a powerful tool in promoting social justice. It can increase access to resources and facilitate communication and collaboration among social justice advocates and organizations. However, there are potential drawbacks to relying too heavily on digital technology to address social justice issues. It is important to ensure that digital technology is designed and implemented in a way that is equitable and inclusive and to avoid the depersonalization of social justice issues. By doing so, we can harness the power of digital technology to promote social justice and create a more equitable world.

The use of digital technology to support mental health is also a topic of great interest and importance. The field of mental health has undergone significant changes in recent years with the growing adoption of digital technology. There is a wide range of digital tools and resources that can be used to support people struggling with mental health issues, such as online counseling, self-help apps, and teletherapy. The benefits of these tools are many and varied. They can offer convenient and accessible support to individuals who may be unable to access traditional mental health services, such as those living in remote or rural areas, or those with mobility issues.

One example of the use of digital technology in mental health is the popular therapy app BetterHelp. BetterHelp provides access to licensed therapists through online chat, phone, and video sessions. The app allows users to easily connect with a therapist and receive personalized care for a variety of mental health concerns.

Furthermore, digital tools can offer a more personalized approach to mental health support. Self-help apps, for example, can provide individuals with tailored resources and strategies based on their unique needs and circumstances. They can also help individuals track their symptoms and progress over time, providing valuable insights into their mental health journey.

There are also potential drawbacks to relying too heavily on digital technology for mental health support. One concern is that these platforms may not be able to provide the same level of human connection and empathy that can be found in face-to-face therapy sessions. The therapeutic alliance, or the relationship between a therapist and client, is an important factor in the success of mental health treatment. While these apps can offer support and guidance, they may not be able to replace the deep human connection that is often necessary for healing and recovery.

Another concern is that digital tools may not be suitable for everyone. Individuals with severe mental health issues may require more intensive and personalized support that cannot be provided through digital means alone. In addition, not everyone has equal access to digital technology, which can exacerbate existing inequalities in mental health outcomes.

Despite these potential drawbacks, digital technology has the potential to play an important role in supporting mental health. By offering convenient, accessible, and personalized support, digital tools can help individuals manage their mental health in a way that works for them. The key is to strike a balance between the benefits of digital technology and the importance of human connection in mental health support. By doing so, we can create a more inclusive and effective approach to mental health care that harnesses the power of technology while also recognizing its limitations.

In conclusion, digital technology has undoubtedly transformed nearly every aspect of our lives. From healthcare and entertainment to transportation and government services, digital technology has revolutionized the way we interact with the world around us. It has brought about unprecedented levels of efficiency, accessibility, and convenience. But, as with any major shift, there are both potential benefits and drawbacks to this technological revolution.

In the realm of healthcare, digital technology has led to significant improvements in patient outcomes and access to care. From telemedicine to electronic health records, digital technology has made it possible for patients to receive care remotely and healthcare professionals to collaborate more effectively. However, the rapid pace of technological change in this field also poses ethical concerns regarding data privacy and security, as well as potential biases in algorithmic decision-making.

The entertainment industry has also been transformed by digital technology. Streaming platforms and social media have made it easier for content creators to reach audiences, and for audiences to access a wider range of content. However, there are concerns regarding the impact of social media on mental health, as well as the potential for algorithmic bias to shape our media consumption.

The transportation industry has seen the growth of ride-sharing services and the potential for autonomous vehicles to revolutionize the way we move around our cities. However, there are concerns about the potential loss of jobs and the impact of autonomous vehicles on public safety.

In the realm of government services, digital technology has led to the growth of e-government services, providing citizens with convenient and accessible services while reducing the costs of service delivery. However, there are concerns regarding data privacy and security, as well as the potential for increased bureaucracy and exclusion of those who lack digital literacy.

Digital technology has also impacted social justice issues, with the potential for increased access to resources and reduced inequality. However, there are concerns about the potential for technology to reinforce existing biases and exclude those who lack access to digital resources.

Finally, digital technology has impacted personal relationships, with the potential for increased connectivity over distance. However, there are concerns regarding the impact of digital technology on the quality of personal relationships and social skills.

As we continue to grapple with the implications of digital technology on our lives, it is important to be mindful of both the potential benefits and drawbacks. We must ensure that technological innovation is driven by a commitment to social responsibility and ethical considerations, while also recognizing the potential for digital technology to improve our lives in significant ways. By embracing these opportunities while remaining vigilant about the risks, we can ensure that digital technology continues to be a force for good in our world.

CHAPTER FOUR: THE FUTURE

The world is constantly changing, and so is the technology that surrounds us. The rapid development of digital transformation has opened up new possibilities and challenges for businesses and individuals alike. From artificial intelligence and machine learning to the internet of things and 5G technology, the future of digital transformation is both exciting and uncertain.

One of the most prominent trends in digital transformation is the increasing use of AI and machine learning. As these technologies become more advanced, they have the potential to revolutionize industries from healthcare to finance. AR and VR are also becoming more mainstream, offering new opportunities for immersive experiences in areas like gaming and education.

Meanwhile, 3D printing is becoming more accessible and affordable, leading to new possibilities in manufacturing and prototyping. Blockchain technology is also gaining traction, offering secure and transparent systems for data storage and transactions.

As more devices become connected through the internet of things, the potential for increased efficiency and productivity grows. However, this also raises concerns about cybersecurity and the protection of sensitive information. Customer experience is also a key area of focus in digital transformation, with businesses leveraging technology to provide personalized and seamless interactions with their customers.

Cloud computing continues to play a significant role in digital transformation, enabling businesses to store and access data from anywhere. And with the advent of 5G technology, the speed and capabilities of wireless networks are poised to increase dramatically.

Finally, the development of quantum computing has the potential to revolutionize computing power, with applications in areas like cryptography and drug discovery. As we look to the future of digital transformation, it's clear that the possibilities are both exciting and complex. It will be up to businesses and individuals to navigate these challenges and leverage the opportunities that arise from this rapidly evolving landscape.

Artificial Intelligence (AI) and Machine Learning (ML) are rapidly advancing technologies that are poised to have a profound impact on the future of business and society. At their core, AI and ML refer to the ability of machines to learn and make decisions based on data, without being explicitly programmed to do so. These technologies are rapidly transforming the way we live and work, and are expected to drive significant advancements across a range of industries.

One area where AI and ML are already having a major impact is in the realm of customer service. Many companies are using chatbots and other AI-powered tools to provide fast, efficient, and personalized customer support. These tools are able to quickly analyze customer inquiries and provide relevant responses, without the need for human intervention. As AI and ML continue to evolve, we can expect to see even more advanced customer service tools, such as virtual assistants that are capable of understanding and responding to complex customer inquiries.

Several companies have been at the forefront of this technological revolution, utilizing and developing AI and ML to gain a competitive advantage in their respective industries.

One of the leading companies in AI and ML is Google. The tech giant has been investing heavily in AI and ML research and development, resulting in the creation of several innovative products and services. Google's AI-powered search engine has become a cornerstone of the company's business, and its machine learning algorithms are used in several other products, such as Google Assistant and Google Maps. In addition, Google has also been working on developing advanced AI systems, such as its DeepMind AI platform, which has been used to solve complex problems in various domains, including healthcare and energy. DeepMind is a British artificial intelligence company that was founded in 2010 by Demis Hassabis, Mustafa Suleyman, and Shane Legg. The company was acquired by Google in 2015 and has since become one of the world's most prominent and innovative AI research organizations.

DeepMind is known for developing cutting-edge AI technologies that are capable of solving complex problems and achieving breakthroughs in various fields such as healthcare, energy, and finance. One of the company's most notable achievements is the creation of AlphaGo, an AI program that defeated the world champion at the board game Go in 2016. This achievement demonstrated the ability of AI to excel in tasks that were previously considered too complex for machines to handle.

Another major area of focus for DeepMind is healthcare. The company has been working on developing AI tools that can assist doctors and nurses in diagnosing and treating patients. For example, DeepMind has partnered with the UK's National Health Service (NHS) to develop an AI system called Streams, which can detect early signs of kidney failure by analyzing patient data in real-time. The system has already been implemented in several hospitals in the UK and has shown promising results in improving patient outcomes.

In addition to healthcare, DeepMind is also working on developing AI technologies that can help address some of the world's most pressing environmental challenges. For example, the company has partnered with energy companies to develop AI systems that can optimize the use of renewable energy sources such as wind and solar power. By analyzing data on weather patterns and energy consumption, these systems can help reduce reliance on fossil fuels and increase the efficiency of renewable energy production.

One of the key features of DeepMind's AI technology is its use of machine learning algorithms. Machine learning is a type of AI that allows machines to learn from data and improve their performance over time without being explicitly programmed. DeepMind has developed several advanced machine learning algorithms that are capable of learning from large datasets and making predictions with a high degree of accuracy.

One of the most notable machine learning algorithms developed by DeepMind is deep learning. Deep learning is a type of machine learning that is inspired by the structure and function of the human brain. Deep learning algorithms consist of multiple layers of artificial neurons that process information in a hierarchical manner. This approach has proven to be highly effective in solving complex problems such as image and speech recognition.

DeepMind's AI technology has also been used to improve the user experience of Google's products and services. For example, the company has developed an AI-powered voice assistant called Google Assistant, which is integrated into devices such as Google Home and Android smartphones. Google Assistant uses natural language processing and machine learning algorithms to understand user queries and provide relevant responses. This has made it easier for users to interact with their devices and access information.

DeepMind is one of the world's leading AI research organizations, known for developing cutting-edge AI technologies that are capable of solving complex problems and achieving breakthroughs in various fields. The company's focus on healthcare, energy, and the environment has the potential to make a significant impact on society, while its use of advanced machine learning algorithms such as deep learning has the potential to revolutionize the field of AI. Through its partnership with Google, DeepMind's technology is already being used to improve the user experience of Google's products and services, and the company's future developments are eagerly anticipated by the AI research community.

Another company that has been at the forefront of AI and ML is Amazon. The e-commerce giant has been using AI and ML algorithms to improve its recommendation engine, which helps customers find products they may be interested in. Amazon has also been developing advanced AI-powered solutions, such as its Alexa virtual assistant, which has become one of the most popular voice assistants in the market. In addition, Amazon has been investing in AI research and development, which has led to the creation of several innovative products, such as the Amazon Go store, which uses AI and ML to provide a cashier-less shopping experience. Amazon Go is a chain of convenience stores that utilizes a range of cutting-edge technologies, including artificial intelligence (AI), to provide a seamless shopping experience for customers. The first Amazon Go store was opened in

Seattle in 2018, and since then, the company has expanded to more than 20 locations in the United States.

One of the key features of Amazon Go stores is their use of AI-powered cameras and sensors to track customer movements and purchases. When a customer enters the store, they scan a QR code on their phone using the Amazon Go app. This code is linked to their Amazon account, and allows the store's cameras and sensors to track the customer's movements and the items they select from the shelves.

As the customer adds items to their basket or removes items, the AI-powered cameras and sensors update the virtual cart in the customer's Amazon Go app. When the customer is finished shopping, they simply walk out of the store - there are no cashiers or checkouts. The items they have selected are automatically charged to their Amazon account, and a receipt is sent to the app.

The use of AI in Amazon Go stores has several benefits for both customers and the company. For customers, the shopping experience is much more convenient and efficient, as there are no queues or checkouts to deal with. For the company, the use of AI enables them to track customer behavior and preferences, which can be used to optimize store layouts and product offerings.

However, there are also concerns around the use of AI in Amazon Go stores, particularly around privacy and job losses. Critics have raised concerns about the amount of data that Amazon is collecting on customers, and how this data might be used in the future. There are also concerns that the widespread adoption of AI in retail could lead to job losses for human cashiers and other retail workers.

Despite these concerns, Amazon Go has been widely praised for its innovative use of AI and its ability to provide a seamless shopping experience for customers. The success of the platform has also inspired other retailers to explore the use of AI and other technologies in their own stores.

Facebook is another company that has been utilizing and developing AI and ML to improve its products and services. The social media giant has been using ML algorithms to personalize the content shown to users on their news feed. In addition, Facebook has been investing in AI research and development, resulting in the creation of several innovative products, such as its DeepFace facial recognition system. Facebook's DeepFace is a facial recognition system developed by the company's AI research division. The system is designed to recognize faces in images and videos with an accuracy rate of 97.35%, which is comparable to human-level performance. This system has been used by Facebook to tag users in photos and improve its automatic photo tagging feature.

DeepFace uses deep learning algorithms, a type of machine learning that uses neural networks with multiple layers to analyze data. The system is trained on a large dataset of images, learning to recognize patterns in facial features and develop a model for identifying faces. The algorithm is then able to analyze new images and determine if they contain a face, and if so, identify the person in the image.

One of the key challenges in developing facial recognition technology is dealing with variations in lighting, pose, and expression. DeepFace is able to handle these challenges by using a 3D model of a face to account for variations in pose and lighting. This allows the system to recognize faces even when they are partially occluded or viewed from an angle.

While DeepFace has been successful in recognizing faces in images and videos, it has also been met with criticism and concerns over privacy and surveillance. Facial recognition technology raises concerns about the potential for misuse and abuse by governments and private companies. There are concerns about the accuracy of the technology, particularly in identifying individuals from minority groups, and the potential for it to be used for discriminatory purposes.

Facebook has acknowledged these concerns and implemented measures to address them. For example, the company allows users to opt-out of facial recognition technology and has put limits on how the technology can be used by third-party developers.

In addition to its use in photo tagging and recognition, Facebook has also applied facial recognition technology to other areas. The company has used it to improve its ad targeting, for example by identifying the gender and age of people in photos. It has also used the technology to identify and remove fake accounts, which is an ongoing issue for social media platforms.

Overall, DeepFace is a powerful tool that has the potential to revolutionize the way we interact with technology. Its ability to accurately recognize faces has numerous applications in fields such as security, marketing, and healthcare. However, its development and use also raise important ethical considerations that must be carefully considered and addressed.

Facebook has also been working on developing AI systems to improve its content moderation process, which has become increasingly important as the platform has come under scrutiny for its role in spreading misinformation and hate speech.

Microsoft is another company that has been investing heavily in AI and ML research and development. The software giant has been using AI and ML algorithms to improve its products, such as its Office suite and Bing search engine. Microsoft has also been working on developing advanced AI-powered solutions, such as its Cortana virtual assistant and its HoloLens augmented reality headset. In addition, Microsoft has been investing in AI research, resulting in the creation of several innovative products, such as its Healthcare NExT initiative, which aims to use AI to improve healthcare outcomes.

Tesla is a company that has been using AI and ML to disrupt the automotive industry. The electric car manufacturer has been using AI and ML algorithms to improve the performance and safety of its vehicles. Tesla's Autopilot system, which uses AI and ML to control the vehicle, has been praised for its advanced features, such as lane centering and automatic emergency braking. In addition, Tesla has been investing in AI research and development, resulting in the creation of several innovative products, such as its Full Self-Driving (FSD) system, which aims to enable fully autonomous driving.

IBM is another company that has been investing heavily in AI and ML research and development. The technology giant has been using AI and ML algorithms to improve its products and services, such as its Watson cognitive computing platform. IBM Watson is a cognitive computing system that was developed by IBM. It is named after the company's founder, Thomas J. Watson, and was designed to provide advanced data analysis and problem-solving capabilities. Watson was first introduced to the public in 2011, when it competed on the game show Jeopardy! against two of the show's greatest champions and won.

Since its debut on Jeopardy!, Watson has become one of the most recognizable and influential artificial intelligence (AI) platforms in the world. The platform uses natural language processing and machine learning to analyze massive amounts of data, understand complex questions and problems, and generate insights that can help organizations make better decisions.

Watson's capabilities have been applied across a wide range of industries, including healthcare, finance, education, and more. In healthcare, Watson has been used to help doctors diagnose diseases, identify treatments, and develop personalized care plans for patients. It has also been used to analyze medical records and clinical data to help hospitals and health systems identify opportunities for improving patient outcomes.

In finance, Watson has been used to analyze financial data and identify trends and patterns that can help traders and investors make better investment decisions. It has also been used to analyze customer data and develop personalized marketing campaigns that are more likely to be effective.

In education, Watson has been used to develop personalized learning plans for students based on their individual strengths and weaknesses. It has also been used to analyze student data and identify opportunities for improving teaching methods and curriculum.

One of the most notable applications of Watson is in the field of artificial intelligence itself. Watson has been used to develop and train other AI systems, including chatbots and virtual assistants. IBM has also made Watson available as a cloud-based service, allowing organizations of all sizes to access its powerful capabilities without the need for expensive hardware and software investments.

Watson has also been at the forefront of innovation in the field of quantum computing. In 2016, IBM announced that it would make Watson available to researchers and developers working on quantum computing, allowing them to use the platform to design and test quantum algorithms.

However, Watson has not been without its challenges and criticisms. In 2018, it was reported that Watson's performance in the healthcare industry had fallen short of expectations, with some doctors questioning the accuracy and usefulness of the platform's recommendations.

Despite these challenges, IBM continues to invest in the development and improvement of Watson, recognizing its potential to transform a wide range of industries and drive innovation in the field of artificial intelligence.

IBM Watson has emerged as one of the most influential and recognizable artificial intelligence platforms in the world, with applications across a wide range of industries. While there have been challenges and criticisms along the way, Watson's capabilities in natural language processing, machine learning, and data analysis make it a powerful tool for organizations looking to make better decisions and drive innovation. As technology continues to evolve and new AI platforms emerge, Watson will undoubtedly continue to be a key player in the future of artificial intelligence.

IBM has also been working on developing advanced AI-powered solutions, such as its Project Debater, which uses AI to engage in debates with humans. In addition, IBM has been investing in AI research, resulting in the creation of several innovative products, such as its AI-powered weather forecasting system.

AI and ML are also being used to drive advancements in healthcare. For example, machine learning algorithms can be used to analyze large datasets of patient information and identify patterns that can help doctors make more accurate diagnoses and treatment decisions. AI-powered tools are also being used to develop new treatments and medications, by simulating the effects of different drug compounds on biological systems.

The future of transportation is also expected to be heavily influenced by AI and ML. Self-driving cars are already being tested on roads around the world, and are expected to become more common in the coming years. These vehicles are powered by complex machine learning algorithms that allow them to navigate roads, avoid obstacles, and respond to changing traffic conditions in real-time. In addition to self-driving cars, AI and ML are also being used to optimize traffic flow and reduce congestion in cities.

While AI and ML have the potential to drive significant advancements across a range of industries, there are also concerns about the impact of these technologies on jobs and the economy. Some experts predict that AI and ML will lead to widespread automation of jobs, particularly in industries such as manufacturing and transportation. While this may lead to increased productivity and efficiency, it could also lead to significant job displacement and economic disruption.

Another concern is the potential for AI and ML to perpetuate existing biases and inequalities. For example, if AI algorithms are trained on biased datasets, they may produce biased results. This could have significant implications in areas such as criminal justice and hiring, where decisions based on biased algorithms could have serious consequences for individuals and society as a whole.

Despite these challenges, there is no denying the potential of AI and ML to drive significant advancements in a range of industries. As these technologies continue to evolve, we can expect to see even more advanced and innovative applications of AI and ML in the years to come. However, it will be important to carefully consider the potential risks and challenges associated with these technologies, and work to ensure that they are developed and deployed in an ethical and responsible manner.

Augmented reality (AR) and virtual reality (VR) are two of the most talked-about technologies in recent years, offering new and exciting ways for people to interact with digital content and the world around them. While they share some similarities, they are distinct technologies with unique strengths and limitations.

AR is a technology that overlays digital information onto the real world, typically through the use of a smartphone or other mobile device. This can include visual information such as text or images, as well as audio or haptic feedback. AR has been used in a variety of applications, from gaming and entertainment to education and training.

One example of successful AR implementation is the Ikea Place app, which allows users to preview furniture in their home before making a purchase. By using the camera on their smartphone, users can place virtual furniture in their real-world environment and see how it looks and fits before making a decision. This not only improves the buying experience for customers but also reduces the number of returns and exchanges for Ikea.

However, AR also has some limitations. One major drawback is that it requires a device such as a smartphone or tablet to access the digital content, which can be a barrier for some users. Additionally, AR experiences can be limited by the quality of the camera and the device's processing power, which can impact the accuracy and realism of the AR overlay.

In contrast, VR is a technology that creates a completely immersive digital environment that users can interact with using a headset and other peripherals. VR has been used in a variety of applications, from gaming and entertainment to education and training.

One example of successful VR implementation is the medical training platform Surgical Theater, which allows medical professionals to practice complex surgical procedures in a virtual environment before performing them on patients. This not only improves patient outcomes by reducing errors and complications but also provides a safe and cost-effective way for medical professionals to train.

However, VR also has some limitations. One major drawback is the cost of the equipment, which can be prohibitively expensive for some users. Additionally, VR experiences can be limited by the quality of the graphics and the processing power of the device, which can impact the realism and immersion of the virtual environment.

While AR and VR have some similarities, they are distinct technologies with unique strengths and limitations. AR is well-suited for applications that require the overlay of digital information onto the real world, while VR is ideal for creating immersive digital environments that users can interact with.

The future of AR and VR is bright, with both technologies expected to play an increasingly important role in a variety of industries. For example, AR is already being used in fields such as construction and manufacturing to provide workers with real-time information and guidance, while VR is being used in fields such as education and therapy to create safe and immersive learning environments.

One potential area of growth for both AR and VR is in the field of remote work and collaboration. With more people working remotely than ever before, AR and VR could provide a way for teams to collaborate and communicate in a more immersive and engaging way. For example, a team working on a design project could use AR to overlay digital models onto the real world and collaborate in real-time, while a team working on a complex problem could use VR to create a shared virtual environment for brainstorming and ideation.

Virtual and Augmented Reality technologies require specialized hardware to deliver an immersive experience to users. These devices have come a long way since the early days of VR headsets that were bulky, uncomfortable, and expensive. Today, a range of headsets and other devices is available to cater to the needs of different users and applications.

VR headsets are the most common device used to experience virtual reality. These headsets are worn on the head and typically cover the eyes, ears, and sometimes the mouth. The aim is to provide an immersive experience by blocking out external stimuli and providing a simulated environment for the user. The most popular VR headsets are the Oculus Rift and the HTC Vive, which are designed for high-end PCs, and the PlayStation VR, which is designed for use with the Sony PlayStation 4 & 5 gaming console. These headsets offer high-quality displays, high refresh rates, and accurate motion tracking, which helps to reduce motion sickness and enhances the overall experience.

Mobile VR headsets are a more affordable alternative to high-end VR headsets. These devices are powered by smartphones and work by inserting the phone into the headset, which uses the phone's display and sensors to create a virtual reality experience. The Samsung Gear VR and Google Daydream View are two popular examples of mobile VR headsets. They offer a more accessible and portable VR experience, but their performance is limited compared to high-end VR headsets.

AR headsets differ from VR headsets in that they allow users to interact with the real world while overlaying digital information. These devices typically consist of a headset with a transparent display that can overlay digital information onto the user's field of view. The most popular AR headset is the Microsoft HoloLens, which is designed for use in professional settings. It allows users to view and interact with digital information, such as schematics or 3D models, while still being able to see the real world.

Smart glasses are a type of AR headset that resemble regular glasses. They typically feature a small display in the corner of the user's field of view and can overlay digital information onto the user's view. One example of smart glasses is the Google Glass, which was discontinued in 2015 but has recently been reintroduced as an enterprise product.

VR and AR hardware have come a long way since their inception, and there are now a range of devices available to cater to different needs and budgets. These devices have the potential to revolutionize how we experience and interact with digital information, and their continued development and improvement will likely lead to further advancements in the field.

AR and VR are two exciting and rapidly evolving technologies that have the potential to transform the way we interact with digital content and the world around us. While they have some limitations, both technologies are expected to play an increasingly important role in a variety of industries in the coming years.

As we dive deeper into the future of digital transformation, it is important that we touch on a highly important topic, 3D Printing. 3D printing is a revolutionary technology that has been gaining popularity over the past decade. Also known as additive manufacturing, 3D printing is a process of creating three-dimensional objects from a digital file. It has been used in various fields, from aerospace to healthcare, and has been an important tool for prototyping and product development.

The history of 3D printing can be traced back to the 1980s when Chuck Hull, the co-founder of 3D Systems, invented the first 3D printing process, called stereolithography. It was initially used for creating small, intricate parts for the automotive and aerospace industries. Over the years, the technology has advanced, and 3D printers are now capable of producing larger and more complex objects.

One of the major advantages of 3D printing is its ability to create customized objects. For example, in healthcare, 3D printing has been used to create prosthetic limbs and implants that are tailored to fit individual patients. This has led to better outcomes and improved quality of life for patients.

Another advantage of 3D printing is its potential to reduce waste and environmental impact. Traditional manufacturing processes often involve creating excess material that is discarded. With 3D printing, only the necessary material is used, reducing waste and minimizing the environmental impact of production.

3D printing has come a long way since its inception, and today, it's used to create an astonishing array of objects that were once considered impossible to produce. From prosthetic limbs and car parts to customized toys and jewelry, 3D printing has made it possible to produce highly complex shapes and designs with great accuracy and speed.

One of the most impressive 3D printed objects to date is the Strati, the world's first 3D printed car. Created by Local Motors in 2014, the Strati is a fully functional electric car that was produced in just 44 hours using a large-scale 3D printer. The car's chassis, body, and even its seats were all printed using a technique known as fused deposition modeling, which involves extruding layers of melted plastic to build up the object layer by layer.

Another remarkable 3D printed object is the Airbus A320neo aircraft engine. Developed by Airbus and GE Aviation, this engine features a 3D printed fuel nozzle that's made of a high-strength titanium alloy. The nozzle was produced using a technique called direct metal laser sintering, which involves melting layers of metal powder with a laser to create the object. The result is a lightweight, durable part that's able to withstand extreme temperatures and pressures.

On a smaller scale, researchers have also used 3D printing to create highly detailed models of human organs, such as the heart and liver. These models are used by medical professionals to plan surgeries and to educate patients about their conditions. By using 3D printing to create these models, doctors can get a better understanding of the patient's anatomy and can plan surgeries with greater precision.

Overall, the possibilities of 3D printing are virtually endless, and as the technology continues to evolve, we can expect to see even more impressive 3D printed objects in the future.

However, 3D printing also has some drawbacks. The technology is still relatively expensive and not widely accessible, which limits its use in certain fields. In addition, the quality of 3D printed objects is not always consistent, and the process can be slow.

In terms of how 3D printing works, the process involves creating a digital model of the object to be printed using computer-aided design (CAD) software. The 3D printer then reads the digital file and creates the object layer by layer, using materials such as plastics, metals, and ceramics.

Looking to the future, 3D printing is expected to continue to evolve and become more widely used. It has the potential to revolutionize manufacturing by allowing for greater customization and reducing waste. In addition, 3D printing could also have implications for space exploration, which we will talk about in a later chapter, as it could allow for the creation of tools and structures in space using local materials.

Several companies, such as Stratasys, 3D Systems, and Ultimaker, have been at the forefront of 3D printing technology and have developed a range of 3D printers for various industries. In addition, many universities and research institutions are also exploring the potential of 3D printing in fields such as medicine and architecture.

Overall, 3D printing is a technology that has the potential to transform manufacturing and create new opportunities for customization and sustainability. As the technology continues to evolve and become more accessible, it will be interesting to see how it is utilized in various fields and what new innovations it will bring.

As the world becomes increasingly connected, the Internet of Things (IoT) has emerged as one of the most significant technological advancements of our time. The IoT refers to the interconnected network of devices, vehicles, appliances, and other physical objects that have the ability to collect and exchange data with each other. This technology has the potential to revolutionize the way we live and work, with its impact expected to be felt across many different industries.

The concept of smart homes is not a distant futuristic dream anymore. It has become a reality that is transforming the way we live. With the help of the Internet of Things (IoT), smart homes are becoming more sophisticated, efficient, and convenient.

Gone are the days when you had to manually adjust your thermostat, lighting, and security systems. With smart home technology, you can control these things from anywhere, using your smartphone or voice commands. Imagine arriving home from a long day at work and walking into a warm, well-lit, and welcoming space, all thanks to the magic of smart home technology.

Smart homes are not just about convenience, though. They can also help you save money on your energy bills by automatically adjusting your heating and cooling systems based on your schedule and preferences. Plus, with smart home security systems, you can have peace of mind knowing that your home is protected around the clock.

One of the most popular examples of smart home technology is the Amazon Echo and Alexa system. It allows you to control a wide range of smart devices using voice commands. For example, you can ask

Alexa to turn on the lights, adjust the thermostat, or even order groceries online.

Another popular smart home technology is the Nest thermostat, which learns your preferences and automatically adjusts your home's temperature based on your habits. With this technology, you can save up to 15% on your energy bills.

Smart home technology is also making life easier for people with disabilities or elderly individuals who may struggle with daily tasks. Smart home devices can be programmed to assist with tasks such as turning on lights, locking doors, or adjusting the temperature, making it possible for these individuals to live more independently.

As the technology behind smart homes continues to advance, the possibilities for how it can enhance our lives are endless. From smart appliances to smart lighting, the future of smart homes is looking brighter than ever before. The convenience and efficiency that smart homes offer are becoming increasingly accessible and affordable, making it easier for more people to enjoy the benefits of a smarter way of living.

The IoT also has the potential to bring about significant changes to various industries such as healthcare, transportation, and manufacturing. In healthcare, IoT devices such as wearable health monitors and medical sensors can help doctors and healthcare providers better monitor patients and provide more personalized treatment. Medical sensors, such as smart pills, also known as digital pills, are a type of medication that contain ingestible sensors that transmit data about the medication to a connected device. This technology is designed to help patients and healthcare professionals track medication adherence and improve treatment outcomes.

The way smart pills work is by incorporating a tiny sensor, typically made of copper, magnesium or other safe materials, into a pill. When the pill is ingested, the sensor is activated by stomach acids and sends a signal to a connected device, such as a smartphone or a wearable device. The device then collects data on medication ingestion, including the time and date of ingestion, and sends it to a healthcare provider.

Smart pills have numerous potential applications, particularly in the treatment of chronic diseases such as diabetes and hypertension, which require careful medication management. By providing real-time data on medication adherence, smart pills can help healthcare professionals adjust treatment plans as needed and ensure that patients are receiving the right dosage of medication at the right time.

One of the leading companies in the development of smart pills is Proteus Digital Health, which has developed a range of digital pills for various medical conditions. For example, Proteus has developed a smart pill for patients with hypertension that includes a medication sensor, a wearable sensor, and a mobile app to track medication adherence and provide feedback to patients.

Another application of smart pills is in the field of clinical trials, where they can be used to track medication adherence and monitor patient outcomes in real time. This can help researchers better understand the efficacy of new medications and improve the design of future clinical trials.

While the potential benefits of smart pills are clear, there are also potential drawbacks and concerns around privacy and data security. Critics argue that the use of smart pills raises ethical questions around the collection and use of patient data, particularly in cases where patients may not fully understand the implications of their data being shared with healthcare providers and third-party companies.

Despite these concerns, the market for smart pills is expected to continue to grow in the coming years as healthcare providers and patients look for new and innovative ways to improve treatment outcomes and better manage chronic diseases. As with any new technology, however, it will be important to strike a balance between the potential benefits and risks of smart pills to ensure that patients and healthcare providers are able to fully leverage this technology while protecting patient privacy and security.

Now in transportation, IoT devices can help to create more efficient and safer roads through the use of connected vehicles and smart traffic systems. In manufacturing, the IoT can help to create more efficient and automated production lines, reducing costs and improving productivity.

The future impact of the IoT on society and business is expected to be significant. As more devices become connected, we will see an increase in the amount of data generated, leading to new insights and opportunities for businesses to improve their operations and create more personalized experiences for their customers. For example, in retail, the IoT can enable stores to create more personalized shopping experiences by tracking customer behavior and preferences through connected devices.

However, with the benefits of the IoT also come potential risks and challenges. The sheer volume of data generated by IoT devices can present privacy and security concerns, as this data may be vulnerable to hacking or misuse. Additionally, the IoT requires significant infrastructure and connectivity, such as 5G, to function properly, which can be a challenge in less developed areas or countries.

Despite these challenges, the IoT is expected to continue to grow and expand in the coming years, with estimates projecting that there will be over 75 billion connected devices by 2025. The IoT has the potential to create a more connected and efficient world, while also presenting new challenges that must be addressed.

As technology continues to evolve, one area that has gained significant attention in recent years is blockchain technology. Blockchain is a distributed ledger technology that allows for secure and transparent transactions without the need for intermediaries. It has the potential to revolutionize the way businesses and individuals conduct transactions and exchange value.

Blockchain technology has already seen significant adoption in the financial industry, with many banks and financial institutions exploring its potential to reduce transaction costs and increase efficiency. But the future of blockchain technology extends far beyond finance. The technology has the potential to transform many other industries, from supply chain management to healthcare to real estate.

Cryptocurrencies have emerged as a new and disruptive force in the financial industry, challenging the traditional models of finance and introducing new possibilities for transactions and investments. While the use of cryptocurrencies is still largely unregulated and their long-term impact on the financial industry is still unclear, there is no denying their potential to change the way we think about money.

One of the most significant ways in which cryptocurrencies have impacted the financial industry is through the introduction of new ways to invest and trade. Cryptocurrencies such as Bitcoin and Ethereum have created a new market for investment, attracting investors from around the world who are interested in the potential returns that these currencies can offer. The decentralized and open nature of cryptocurrency markets has also allowed for greater access and participation by individuals who were previously excluded from traditional financial markets.

In addition to investments, cryptocurrencies have also introduced new possibilities for transactions and payments. Cryptocurrencies can be used to purchase goods and services, and can be transferred from one person to another without the need for intermediaries such as banks or payment processors. This has the potential to significantly reduce the cost and complexity of financial transactions, making it easier and more affordable for individuals and businesses to move money around the world.

As cryptocurrencies have become increasingly popular, so have the number of crypto exchanges and wallets available for users to store, trade and manage their digital assets. A cryptocurrency exchange is a platform where users can buy, sell, and trade different cryptocurrencies, whereas a cryptocurrency wallet is a digital wallet used to store and manage cryptocurrencies.

Crypto exchanges are available in various forms such as centralized, decentralized, and peer-to-peer exchanges. Centralized exchanges are the most common and allow users to trade cryptocurrencies on a platform that is controlled by a central authority. Decentralized exchanges, on the other hand, allow users to trade cryptocurrencies without a central authority. These exchanges operate on a blockchain network, and trades are facilitated by smart contracts. Peer-to-peer exchanges, as the name suggests, allow users to directly buy or sell cryptocurrencies with one another without the need for an intermediary.

Some of the most popular centralized crypto exchanges include Binance, Coinbase, Kraken, and Bitfinex. These exchanges allow users to buy, sell and trade a variety of cryptocurrencies, including Bitcoin, Ethereum, and Litecoin. Centralized exchanges typically charge fees for trading and may also have additional fees for withdrawals or deposits.

In addition to exchanges, cryptocurrency wallets are also essential for managing digital assets. Crypto wallets can be hardware devices, software applications, or online services that allow users to securely store and manage their digital assets. Hardware wallets such as Ledger and Trezor provide an added layer of security as they are offline and require physical access to be used. Software wallets such as Exodus and MyEtherWallet are more convenient as they are accessible through a computer or mobile device.

As with any digital platform, there are potential risks associated with crypto exchanges and wallets. For example, centralized exchanges can be vulnerable to hacking or other security breaches, resulting in the loss of user funds. Additionally, users may be at risk of losing their digital assets if they lose access to their private keys or if they are stolen. Therefore, it is important for users to take precautions such as using two-factor authentication, choosing a reputable exchange or wallet provider, and keeping their private keys secure.

The rise of cryptocurrencies has also created new challenges for the financial industry. The unregulated nature of these currencies has made them a target for fraud and illegal activity, and their decentralized nature has made it difficult for regulators to monitor and control the market. In addition, the high volatility of cryptocurrencies has made them a risky investment for some, with prices fluctuating rapidly and unpredictably.

Despite these challenges, the financial industry has recognized the potential of cryptocurrencies and blockchain technology, and many companies are investing in their development and integration into traditional financial systems. Major financial institutions such as JPMorgan and Goldman Sachs have developed their own blockchain-based systems for trading and settlement, and many others are exploring the possibilities of cryptocurrencies and blockchain for everything from identity verification to cross-border payments.

Overall, the future of cryptocurrencies and blockchain technology in the financial industry is still uncertain, but it is clear that these technologies have already had a significant impact and will continue to shape the future of finance. As the market for cryptocurrencies and blockchain matures, it will be important for regulators, investors, and businesses to work together to ensure that these technologies are used responsibly and sustainably for the benefit of all.

One of the most promising applications of blockchain technology is in supply chain management. By creating a secure and transparent ledger of all transactions and movements within a supply chain, blockchain technology can improve efficiency, reduce fraud, and increase transparency. By recording the entire history of a product's journey from manufacturer to end-user on a blockchain ledger, consumers can verify the authenticity and quality of the product and ensure that it has been produced and transported in an ethical and sustainable manner. This can help to build consumer trust, increase brand loyalty, and reduce the risks of product recalls, counterfeiting, and other supply chain-related issues.

VeChain is a blockchain-based platform that was created to improve supply chain management and business processes. It aims to create a transparent, traceable, and efficient supply chain ecosystem by integrating blockchain technology, the Internet of Things (IoT), and artificial intelligence (AI) technologies. The VeChain platform is designed to help businesses optimize their operations, reduce costs, and increase efficiency by providing a trusted and secure platform for data sharing.

One of the key features of VeChain is its ability to provide users with real-time data on the status of their products throughout the supply chain. This allows businesses to monitor their products from the moment they are produced to the moment they are delivered to the end customer. VeChain's platform uses sensors and other IoT devices to collect data on the products as they move through the supply chain, and this data is then recorded on the blockchain in a tamper-proof and transparent manner.

Another important feature of VeChain is its ability to provide businesses with a way to track and verify the authenticity of their products. The platform uses a unique ID system called VeChainThor ID, which assigns a unique digital ID to each product on the platform. This ID is then recorded on the blockchain, providing an immutable record of the product's authenticity and provenance. This is particularly important for high-value goods such as luxury items, fine art, and pharmaceuticals, where counterfeiting is a major problem.

VeChain is already being used by several large companies, including Walmart China, BMW, and H&M. Walmart China, for example, is using VeChain to track the supply chain of its food products, including pork and rice. The platform provides Walmart China with real-time data on the quality and safety of its products, allowing the company to quickly identify and address any issues that arise.

In addition to its use in supply chain management, VeChain is also being used in other industries such as healthcare and finance. In the healthcare industry, VeChain is being used to create a digital health passport that stores an individual's medical history on the blockchain. This allows healthcare providers to access a patient's medical history quickly and easily, improving the quality of care they receive. In the finance industry, VeChain is being used to create a secure and transparent platform for trade finance, reducing the risk of fraud and improving the efficiency of the trade finance process.

Overall, VeChain is a promising blockchain platform that has the potential to revolutionize supply chain management and other industries. Its use of IoT devices and AI technologies, combined with its focus on transparency and trust, make it a powerful tool for businesses looking to improve their operations and gain a competitive edge in the market.

Several companies are already using blockchain technology to improve their supply chain operations. For example, IBM's Food Trust platform uses blockchain to enable food retailers and producers to track the entire journey of a product from farm to table, ensuring transparency and accountability throughout the supply chain. As mentioned earlier, Walmart, one of the largest retailers in the world, is also using blockchain to track the provenance of its food products, with the aim of reducing waste, improving product safety, and increasing efficiency.

In the pharmaceutical industry, blockchain technology is being used to track the provenance and authenticity of drugs, reducing the risks of counterfeit drugs entering the supply chain and helping to ensure patient safety. For example, the MediLedger project, a collaboration between several pharmaceutical companies, uses blockchain technology to track the movement of drugs from manufacturer to distributor to pharmacy, ensuring that they are genuine and have not been tampered with.

Another potential application of blockchain technology is in healthcare. By using blockchain to securely store and share patient medical records, healthcare providers can improve patient care and reduce costs. Patients would have complete control over their medical records, allowing them to easily share their information with other healthcare providers as needed. This would eliminate the need for multiple records and reduce the risk of errors or omissions.

Blockchain technology also has the potential to transform the real estate industry. By creating a secure and transparent ledger of all property transactions, blockchain technology can reduce fraud, increase transparency, and improve the efficiency of the buying and selling process. In fact, several startups are already working on blockchain-based real estate platforms that allow for faster and more secure property transactions.

Real estate is a massive industry, and it's no surprise that it's been affected by the rise of cryptocurrencies. Real estate cryptocurrencies, or "tokenized real estate," are a way to use blockchain technology to invest in and manage real estate properties.

The concept is relatively straightforward: real estate properties are broken down into "tokens," or fractions of the whole property, which can then be bought and sold on cryptocurrency exchanges. This allows for smaller investors to get involved in real estate investments that they might not have been able to participate in before.

One example of a company utilizing this concept is the startup Harbor, which recently raised $28 million to fund its efforts to tokenize real estate assets. The company is partnering with a variety of real estate firms to launch tokenized real estate projects, which they hope will make investing in real estate more accessible and liquid.

There are also existing companies that are beginning to embrace the use of cryptocurrencies in real estate transactions. For example, the RE/MAX London estate agency recently announced that it would be accepting Bitcoin as a form of payment for its properties.

Not to mention, the metaverse, which is a term used to describe a fully immersive digital world where people can interact with each other and digital objects in a virtual space. It is a concept that has gained a lot of attention in recent years, particularly with the rise of virtual reality and blockchain technology.

Blockchain technology has the potential to play a significant role in the development of the metaverse. One of the key advantages of blockchain technology is its ability to create decentralized and trustless systems. This means that transactions can be recorded on a secure and transparent ledger that is maintained by a network of users, rather than a centralized authority.

This has significant implications for the metaverse, where users will need to be able to trust the digital assets and transactions they engage with. Blockchain technology can help to create a secure and transparent system for managing these assets, allowing users to buy, sell, and trade digital assets with confidence.

There are already a number of blockchain-based platforms that are exploring the potential of the metaverse. For example, Decentraland is a blockchain-based virtual world where users can buy and sell virtual land, build on that land, and interact with other users. The virtual land is represented as non-fungible tokens (NFTs) on the Ethereum blockchain, which means that users can own and transfer their virtual land just like they would with physical property.

Another example is Somnium Space, which is a virtual world that is built on the Ethereum blockchain. Users can buy and sell virtual land, build on that land, and interact with other users in a fully immersive 3D environment. The virtual land is also represented as NFTs on the Ethereum blockchain, which means that users can own and transfer their virtual land.

The use of blockchain technology in the metaverse also has implications for digital identity. In a fully immersive digital world, users will need a way to prove their identity and establish trust with other users. Blockchain technology can help to create a secure and decentralized system for managing digital identity, allowing users to control their own identity and maintain their privacy.

Overall, the combination of blockchain technology, real estate and the metaverse has the potential to revolutionize the way we interact with digital assets and each other in virtual spaces. While there are still many challenges to overcome, the future looks bright for this exciting intersection of technology and imagination.

However, there are still several challenges that need to be addressed before real estate cryptocurrencies become more widespread. One of the biggest issues is the lack of standardization and regulation in the industry. Many countries are still trying to figure out how to regulate cryptocurrency transactions, and the lack of clarity can make it difficult for investors to navigate the market.

Additionally, the process of tokenizing real estate properties can be complicated and expensive, which can make it difficult for smaller companies to get involved. And because these tokens are based on blockchain technology, there are concerns about the security and privacy of these transactions.

Despite these challenges, many experts believe that real estate cryptocurrencies have the potential to revolutionize the industry. By making real estate investments more accessible to a wider range of investors, these tokens could increase liquidity in the market and provide new opportunities for growth.

The future of real estate cryptocurrencies is still uncertain, but it's clear that the industry is ripe for disruption. As more companies begin to experiment with tokenizing real estate assets, it will be interesting to see how the market evolves and whether or not these tokens become a common way to invest in real estate.

Overall, the future of blockchain technology is full of potential. As the technology continues to evolve, we can expect to see even more innovative applications that transform the way we do business and live our lives. It has the potential to revolutionize many industries, including finance, healthcare, and supply chain management. But, as with any new technology, it also raises questions about its implications for society. One such question is whether blockchain technology, combined with other emerging technologies like ISO 20022 and the Great Reset, could lead to a dystopian future.

ISO 20022 is a new global standard for financial messaging that aims to streamline communication between financial institutions. The standard uses a common language and a standardized format for financial messages, making it easier for banks to communicate with each other, even if they are using different systems. ISO 20022 has been developed by SWIFT, the global provider of secure financial messaging services, in collaboration with over 500 financial institutions worldwide.

ISO 20022 has many benefits, including improved data quality, greater efficiency, and increased interoperability between financial systems. However, some experts believe that ISO 20022 could be used to create a global financial surveillance system that could be used to monitor every financial transaction in the world.

The Great Reset is a term coined by the World Economic Forum (WEF) to describe a series of changes that are needed to address the social, economic, and environmental challenges facing the world today. The WEF argues that the current system is unsustainable and that a new approach is needed to create a more equitable and sustainable world.

The Great Reset has many supporters, but it has also been criticized by some who see it as a plan for a new world order. Critics argue that the Great Reset is a way for the global elite to consolidate power and control over the world's resources, including financial resources. They argue that the Great Reset could lead to a dystopian future in which individual freedoms are curtailed, and people are forced to conform to a new set of rules.

So, what is the connection between blockchain technology, ISO 20022, and the Great Reset? The answer lies in the potential for these technologies to be used together to create a new global financial system that could be used to monitor and control every financial transaction in the world.

Imagine a world in which every financial transaction is recorded on a blockchain using the ISO 20022 standard. This would create a vast database of financial information that could be used to monitor and analyze every financial transaction in the world. This information could be used to identify individuals and organizations that are engaging in illegal activities, such as money laundering, terrorist financing, and tax evasion.

Proponents of this system argue that it would create a more secure and transparent financial system that would be less prone to fraud and corruption. However, critics argue that it could lead to a dystopian future in which individual privacy is eroded, and people are forced to conform to a new set of rules.

Blockchain technology, ISO 20022, and the Great Reset are all emerging technologies that have the potential to change the world we live in. However, the combination of these technologies raises questions about their implications for society.

While there are still challenges to overcome, such as scalability, ethical and regulatory hurdles, the benefits of blockchain technology are clear, and it is an exciting time to be part of this technological revolution.

As digital technology continues to transform the way we live and work, the importance of cybersecurity has become more critical than ever. With increasing amounts of sensitive data being stored and transmitted digitally, the risks of cyberattacks and data breaches have also risen significantly. In fact, according to a report by Cybersecurity Ventures, cybercrime is projected to cost the world $10.5 trillion annually by 2025.

The potential impact of a cyberattack on a business or organization can be devastating. In addition to the loss of valuable data, it can also result in financial losses, damage to reputation, and even legal consequences. This is why it is essential for businesses and organizations to have strong cybersecurity measures and protocols in place to protect their data and systems.

One of the main challenges of cybersecurity is the constantly evolving nature of cyber threats. Hackers and cybercriminals are always looking for new ways to exploit vulnerabilities and gain access to sensitive data. This is where the use of advanced technologies such as artificial intelligence and machine learning comes into play.

By leveraging these technologies, businesses and organizations can develop more sophisticated cybersecurity measures that can detect and respond to threats in real-time. For example, machine learning algorithms can analyze vast amounts of data to identify patterns and anomalies that may indicate a potential cyber threat.

Another critical aspect of cybersecurity is the need for strong encryption protocols to protect data that is transmitted over networks. This is where blockchain technology can be particularly useful. By using decentralized networks and advanced cryptographic algorithms, blockchain technology can provide a high level of security for sensitive data.

In addition, cloud-based security solutions are becoming increasingly popular as more businesses move their operations to the cloud. Cloud-based security solutions offer a range of benefits, including improved scalability, flexibility, and cost-effectiveness. These solutions can also provide real-time threat detection and response, as well as continuous monitoring and compliance reporting.

One of the most critical cybersecurity technologies is encryption. Encryption is the process of converting data into a code that can only be accessed with a decryption key. This technology helps to protect sensitive data from cyber attackers and ensure that it remains secure during transmission and storage.

Finally, multi-factor authentication (MFA) is another key cybersecurity technology that can help prevent unauthorized access to sensitive data and systems. MFA requires users to provide multiple forms of identification, such as a password and a fingerprint scan, before accessing a system or application. This added layer of security can help prevent cybercriminals from gaining access to sensitive data and systems.

However, despite the potential benefits of these technologies, there is also the risk of relying too heavily on them. It is important to recognize that no technology is foolproof, and even the most advanced cybersecurity measures can be breached. This is why it is essential to maintain a multi-layered approach to cybersecurity, incorporating a range of technologies, protocols, and best practices.

Going into the future, it is clear that cybersecurity will continue to be a critical issue for businesses and organizations across all sectors. As technology continues to advance and cyber threats become more sophisticated, the need for strong cybersecurity measures and protocols will only become more critical. By staying up-to-date with the latest developments in cybersecurity technology and best practices, businesses and organizations can help to protect themselves and their customers from the potentially devastating consequences of cyberattacks and data breaches.

As we have seen, digital transformation is constantly evolving and advancing, and one of the technologies that is poised to have a significant impact on this field is 5G technology. 5G is the latest generation of mobile network technology, offering higher speeds, lower latency, and greater bandwidth than its predecessors. These improvements have the potential to revolutionize the way we connect and communicate, opening up new possibilities for businesses and individuals alike.

One of the key advantages of 5G technology is its increased speed. With 5G, mobile internet speeds are expected to be up to 100 times faster than 4G, which will enable businesses to communicate and transfer data more quickly and efficiently. This increased speed will also enable the development of new technologies and applications that were not previously possible. For example, the low latency of 5G networks will make it possible to develop real-time virtual reality and augmented reality experiences, which could have a significant impact on industries such as gaming and entertainment.

In addition to speed and latency improvements, 5G technology is also expected to offer greater capacity and connectivity. The increased bandwidth of 5G networks will enable more devices to be connected simultaneously, allowing for the development of smart cities and the Internet of Things (IoT). This increased connectivity will also enable businesses to collect and analyze more data in real-time, providing valuable insights into customer behavior and preferences.

Despite these potential benefits, there are also challenges that need to be addressed. One of the key challenges is the need for significant investment in infrastructure to support the rollout of 5G technology. This includes the installation of new base stations and the upgrading of existing ones, as well as the installation of fiber-optic cables to support the increased capacity of 5G networks. There are also concerns around the security of 5G networks, particularly in light of the increased use of connected devices and the potential for cyber-attacks.

The potential impact of 5G technology on digital transformation is significant. It has the potential to transform the way we connect and communicate, enabling new technologies and applications that were previously impossible. However, there are also challenges that need to be addressed, and it will require significant investment and effort to ensure that the full potential of 5G is realized.

As technology continues to revolutionize the way businesses operate, one of the most exciting developments on the horizon is quantum computing. Quantum computing is a new type of computing technology that uses quantum-mechanical phenomena, such as superposition and entanglement, to perform operations on data. These quantum computers are much faster and more powerful than classical computers, making them potentially useful for a wide range of applications.to perform complex calculations and solve problems that would be impossible with classical computing. The potential benefits of quantum computing are immense, with the ability to revolutionize a wide range of industries, from finance and healthcare to manufacturing and logistics.

One of the key advantages of quantum computing is its ability to process vast amounts of data quickly and efficiently. This is because quantum computers use quantum bits, or qubits, which can exist in multiple states at once. This means that a quantum computer can perform multiple calculations simultaneously, rather than processing data one bit at a time like a classical computer. As a result, quantum computers have the potential to revolutionize the way we process and analyze data, enabling us to tackle complex problems and make more informed decisions.

Another potential benefit of quantum computing is its impact on encryption. As the amount of data we generate and store continues to grow, the need for secure encryption becomes increasingly important. Quantum computers have the potential to break many of the encryption methods that are currently in use, which could have significant implications for the security of sensitive data. However, quantum computing also offers the potential to develop new encryption methods that are more secure and robust than those currently in use.

Despite the potential benefits of quantum computing, there are also significant challenges to be addressed. One of the biggest challenges is developing the hardware and software needed to support quantum computing. The technology is still in its early stages, and researchers are still working to develop the building blocks needed to build a scalable quantum computer. In addition, quantum computing is a highly specialized field that requires expertise in both quantum mechanics and computer science, which means that there are relatively few experts in the field.

Quantum mechanics is a branch of physics that deals with the behavior of matter and energy at the atomic and subatomic level. It was developed in the early 20th century to explain the strange and counterintuitive behavior of particles at this scale. One of the most famous examples of quantum mechanics is the wave-particle duality, which states that particles can behave as both waves and particles simultaneously.

Another important aspect of quantum mechanics is quantum cryptography. This technology uses the principles of quantum mechanics to create secure communication channels that cannot be intercepted or eavesdropped upon. This is because the act of observing quantum particles changes their state, meaning that any attempt to intercept the communication would be immediately detected.

Quantum mechanics has also been used to develop new types of sensors, such as atomic clocks and quantum magnetometers, that are much more precise and accurate than traditional sensors. These sensors have important applications in fields such as navigation, telecommunications, and environmental monitoring.

Another challenge is the potential impact of quantum computing on existing industries and job markets. As with any new technology, there is the potential for disruption as industries and job roles are transformed. While quantum computing is still in its early stages, it is likely that the technology will have a significant impact on a wide range of industries, from finance and healthcare to logistics and manufacturing. As a result, it will be important for businesses and governments to develop strategies for adapting to these changes and minimizing any potential negative impacts.

The potential impact of quantum computing on digital transformation is immense, with the ability to revolutionize the way we process and analyze data, as well as develop new and more secure encryption methods. However, there are also significant challenges to be addressed, including the development of the hardware and software needed to support quantum computing, as well as the potential impact on existing industries and job markets. As with any new technology, it will be important for businesses and governments to approach quantum computing with a thoughtful and strategic mindset, in order to realize its full potential while minimizing any potential negative impacts.

In conclusion, the future of digital transformation is a vast and ever-evolving landscape that promises to revolutionize the way we live, work, and interact with the world around us. From the powerful and groundbreaking potential of artificial intelligence and machine learning to the immersive experiences of augmented and virtual reality, the possibilities are endless. The growth of 3D printing and the widespread use of the Internet of Things will create a new era of innovation and convenience, while the groundbreaking potential of blockchain technology and cryptocurrencies offers exciting new opportunities for secure and decentralized transactions.

However, these technological advancements also come with inherent risks, especially in the realm of cybersecurity. As we move towards the future, the threat of data breaches and cyberattacks will continue to grow, and the need for robust cybersecurity measures and protocols will become more critical than ever. The exciting potential of quantum computing offers tremendous benefits for businesses, but it also presents new challenges in data processing and encryption that must be addressed.

In short, the future of digital transformation is both exciting and complex. It requires us to balance the benefits of innovation with the risks of vulnerability and to navigate the challenges of a rapidly changing technological landscape. The key to success in this new era will be to embrace innovation while remaining vigilant and adaptable in the face of emerging threats. By doing so, we can unlock the full potential of digital transformation and create a brighter, more connected future for us all.

CHAPTER FIVE: THE DISTANT, BUT NOT SO DISTANT FUTURE

Welcome to a journey into the future. A journey where we explore the possibilities and potentials of what lies ahead in the ever-evolving world of technology and digital transformation. This is the world of futurology, where we gaze into the crystal ball of possibilities to anticipate and understand the implications of the next wave of technological advancements.

As digital transformation continues to shape and redefine the world we live in, futurology becomes increasingly relevant as we seek to predict and prepare for what lies ahead. With the rapid pace of technological advancement, it's important to understand what the future holds, and the impact it will have on businesses, individuals, and society as a whole. In this chapter, we will explore the concept of futurology and its relevance to digital transformation.

Futurology, also known as futures studies, is the study of possible futures and the potential consequences of present-day decisions. It is a multidisciplinary field that draws on a wide range of disciplines, including economics, sociology, technology, and psychology. The goal of futurology is not to predict the future with certainty, but rather to identify possible scenarios and assess the likelihood of each one.

The roots of futurology can be traced back to the early 20th century, with the publication of works such as H.G. Wells' "The Shape of Things to Come" and J.B.S. Haldane's "Daedalus; or, Science and the Future". In the decades that followed, the field grew in popularity and prominence, with the establishment of organizations such as the World Futures Society and the publication of influential books such as Alvin Toffler's "Future Shock" and "The Third Wave".

In the context of digital transformation, futurology plays a critical role in helping organizations and individuals anticipate and prepare for the changes that are coming. By examining trends and developments in technology, society, and the economy, futurists can identify potential opportunities and challenges, and help organizations develop strategies to capitalize on them.

As we move further into the digital age, the importance of futurology in shaping the future of digital transformation will only grow. The pace of technological change shows no signs of slowing down, and the impact of digital transformation on society and the economy is likely to be significant. By embracing the insights of futurists and incorporating them into their planning and decision-making processes, organizations can stay ahead of the curve and remain competitive in an increasingly digital world.

In the following sections, we will explore some of the key trends and developments in digital transformation, and examine the ways in which futurology can help us understand and prepare for the changes that are coming. From space tourism and self-replicating robots to nanotechnology and the smart cities, we will explore the ways in which these emerging technologies are likely to shape the future, and the steps that organizations can take to stay ahead of the curve.

Now before we get into the super exciting and deep topic of futurology, it's important that we lay the foundation by highlighting some of the topics we touched on before. In order to understand where we are truly going, we must use data to our advantage. In today's world, we are generating an enormous amount of data every second, and this data can be harnessed to make informed predictions about future trends and patterns. This is where big data comes into play.

Big data refers to the vast amount of structured and unstructured data that is generated through various sources such as social media, e-commerce platforms, and IoT devices. The analysis of this data can provide valuable insights into consumer behavior, market trends, and industry patterns, which can be used to predict future outcomes.

For instance, big data analysis played a significant role in predicting the outcome of the 2016 US Presidential election. Analyzing social media conversations and search patterns of voters helped in predicting the eventual outcome of the election. Similarly, the use of big data in the healthcare industry can help predict disease outbreaks and monitor patient health in real-time.

The impact of big data on futurology and digital transformation is immense. By harnessing the power of big data, businesses can predict future trends and patterns, identify potential risks and opportunities, and make informed decisions. In the field of digital transformation, big data can help organizations streamline their operations, improve customer experience, and gain a competitive edge.

However, the use of big data also raises concerns about data privacy and security. As the amount of data being generated continues to grow, it is crucial to ensure that this data is handled ethically and with utmost care. The implementation of data protection laws and regulations can help prevent data breaches and ensure that sensitive information is not misused.

Now as we continue to explore the intersection of futurology and digital transformation, it is crucial to acknowledge the impact of emerging technologies. Emerging technologies refer to innovations that are currently in the early stages of development but have the potential to significantly impact our lives in the future. The rise of emerging technologies such as quantum computing, augmented reality, and nanotechnology are expected to revolutionize the future of digital transformation.

As mentioned in the previous chapter, quantum computing, for example, is a type of computing that uses quantum-mechanical phenomena such as superposition and entanglement to perform operations on data. Quantum computing has the potential to process data at an unprecedented speed, which can significantly impact the fields of medicine, finance, and logistics, among others. Another emerging technology is augmented reality, which is transforming the way we interact with digital information by overlaying computer-generated information on top of the real world. Augmented reality has been used in various industries such as gaming, education, and marketing. Nanotechnology, on the other hand, involves manipulating matter on a molecular and atomic scale. Nanotechnology has the potential to revolutionize various industries such as medicine, electronics, and energy.

The impact of emerging technologies on society and business is undeniable. These technologies have the potential to disrupt traditional industries, create new opportunities, and change the way we interact with the world around us. As we continue to advance in these emerging technologies, it is essential to consider the potential ethical, social, and economic implications they may bring. The future of digital transformation is not only about technological advancements but also about how these advancements can benefit society in a responsible and sustainable way.

In the next sections, we will delve deeper into the potential impact of these emerging technologies on the future of digital transformation. We will examine real-world examples of how these technologies are being used today, if they even exist yet, and the potential impact they may have on society and business in the future.

If you don't mind, let us explore the potential of brain-machine interfaces and neural implants, discuss their current state of development, potential applications, and ethical and social implications.

Brain-machine interfaces (BMIs) are devices that allow a direct connection between the brain and an external device or machine. This technology has been in development for several decades, and recent advancements have led to a surge of interest and investment in the field. Neural implants, on the other hand, are devices that are surgically implanted in the brain to monitor or stimulate neural activity. These devices have a range of potential applications, from treating neurological disorders to enhancing cognitive function.

One of the most promising applications of BMIs and neural implants is in the field of healthcare. For example, BMIs have been used to help patients with paralysis to control robotic prosthetics, enabling them to perform basic tasks and regain some independence. Neural implants have also been used to treat epilepsy, Parkinson's disease, and other neurological disorders, by providing deep brain stimulation that helps to regulate neural activity.

Beyond healthcare, BMIs and neural implants have the potential to revolutionize transportation and entertainment. BMIs could be used to control autonomous vehicles, for example, allowing users to simply think about where they want to go, rather than physically operating the vehicle. In the field of entertainment, BMIs could enable users to experience immersive virtual reality environments by stimulating the brain to produce sensory experiences. Think of the Black Mirror television series.

"Black Mirror" is a British science fiction anthology television series that explores the dark and twisted side of modern technology and its effects on society. Several episodes of the show feature themes related to brain-machine interfaces (BMIs) and neural implants.

One notable episode is "The Entire History of You" from the first season, which centers around a device called a "grain" that is implanted behind a person's ear and records everything they see and hear. The episode explores the consequences of being able to record and replay every moment of one's life, including the impact on personal relationships and mental health.

Another episode, "Playtest" from the third season, follows a man who participates in a virtual reality horror game that uses a neural implant to read his fears and create customized scares. The episode explores the dangers of blurring the lines between reality and virtual reality and the potential risks of neural implants.

Perhaps the most memorable episode related to BMIs and neural implants is "Black Museum" from the fourth season, which features several interconnected stories related to technology and its impact on humanity. One of the stories centers around a doctor who creates a neural implant that allows a dying woman to experience the pain of others, leading to a shocking twist ending that highlights the potential ethical and social implications of such technology.

Overall, the "Black Mirror" series presents a cautionary tale about the potential dangers of technology, particularly in the realm of brain-machine interfaces and neural implants. These episodes serve as a reminder that while these technologies may hold great promise, their development and implementation must be approached with care and consideration for their potential impact on society.

Now, there are several companies that are currently working on developing brain-machine interfaces (BMIs) and neural implants. One such company is Neuralink, founded by Elon Musk in 2016. Neuralink's goal is to create a high-bandwidth, implantable neural interface that can connect human brains directly to computers and other devices. The company's ultimate vision is to use this technology to enable humans to merge with artificial intelligence and achieve symbiosis with machines. This symbiosis with machines is also called transhumanism and can be classified as a controversial topic.

Transhumanism is a movement that advocates for the use of technology to enhance human abilities and capacities. It is founded on the belief that human beings can transcend the limitations of their biological bodies and minds by using science and technology to augment themselves. The ultimate goal of transhumanism is to create a post-human society in which individuals are free to shape their own identities, and to overcome the limitations imposed on them by their biology.

Transhumanism has its roots in the philosophy of humanism, which places a high value on the dignity and worth of the individual. However, while humanism celebrates the inherent worth of human beings, it does not provide a clear path for individuals to reach their full potential. Transhumanism seeks to address this gap by advocating for the use of technology to enhance human abilities and expand the limits of what it means to be human.

One of the main areas of focus for transhumanism is the use of technology to enhance human intelligence. This includes the development of cognitive enhancers such as nootropics and brain-computer interfaces, which can improve memory, attention, and learning ability. These technologies have the potential to transform education and training, and to unlock the full potential of human creativity and innovation.

Another area of interest for transhumanists is the use of technology to enhance physical abilities. This includes the development of prosthetics, exoskeletons, and other devices that can improve strength, agility, and endurance. These technologies can be used to improve the quality of life for individuals with disabilities, as well as to enhance the performance of athletes, soldiers, and emergency responders.

Transhumanism also explores the potential for technology to extend human lifespan and to overcome the limitations of aging. This includes the development of anti-aging therapies and life extension technologies, such as cryonics and mind uploading. These technologies have the potential to transform our understanding of what it means to be mortal, and to provide new opportunities for individuals to pursue their goals and dreams.

While transhumanism offers many potential benefits, it also raises important ethical and social questions. One concern is that the use of technology to enhance human abilities could exacerbate existing inequalities, creating a world in which only the wealthy and privileged have access to the most advanced technologies. Another concern is that the pursuit of transhumanist goals could lead to the creation of a new class of individuals who are fundamentally different from traditional human beings, leading to social, legal, and moral conflicts.

Despite these concerns, transhumanism continues to gain traction as a movement that is reshaping our understanding of what it means to be human. As technology continues to advance at an unprecedented rate, it is likely that transhumanist ideas will play an increasingly important role in shaping our future. Whether these ideas lead to a world of infinite possibility or to new forms of inequality and conflict remains to be seen.

Now, another company working on BMIs and neural implants is Kernel, founded by Bryan Johnson in 2016. Kernel's mission is to develop and commercialize technologies that can measurably expand human cognition and improve human intelligence. The company is focused on creating non-invasive, implantable devices that can interact with the brain to enhance human abilities, such as memory, learning, and decision-making.

In addition to Neuralink and Kernel, there are several other companies that are developing BMIs and neural implants for a variety of applications. For example, Blackrock Microsystems is working on implantable devices that can help people with paralysis to control prosthetic limbs using their thoughts. Synchron, another company in this space, is developing a neural interface that can enable people with paralysis to control a computer or other electronic device using their thoughts.

These companies are all pushing the boundaries of what is possible with BMIs and neural implants, and their work has the potential to revolutionize healthcare, transportation, entertainment, and other industries. While there are certainly challenges and ethical considerations that must be addressed, the promise of these technologies is immense, and they represent an exciting frontier in the field of digital transformation.

While the potential of BMIs and neural implants is vast, there are still many challenges that need to be addressed before these technologies can become mainstream. The current state of research and development is focused on improving the accuracy and reliability of these devices, as well as making them more accessible and affordable for the general public. Additionally, there are significant ethical and social implications to consider, such as issues of privacy and consent, as well as concerns about the potential for misuse or abuse of these technologies.

As you can see, brain-machine interfaces and neural implants represent some of the most exciting and promising technologies on the horizon. With the potential to revolutionize healthcare, transportation, and entertainment, these devices have the power to change the way we interact with the world around us. However, it is important to approach these technologies with caution, taking into account the ethical and social implications that come with such powerful tools. By doing so, we can ensure that these innovations are used in a responsible and beneficial manner, for the betterment of society as a whole.

Ready to read something that is interesting, yet scary? Here we come across the fascinating concept of self-replicating robots. Imagine a world where robots can reproduce themselves without human intervention, leading to exponential growth in their numbers and capabilities. This may sound like science fiction, but the development of self-replicating robots is already underway, and its potential impact on industries such as manufacturing, space exploration, and environmental sustainability is immense.

Self-replicating robots, also known as self-replicating machines or von Neumann machines, are robots that can create copies of themselves without human intervention. The concept was first introduced by mathematician John von Neumann in the 1940s, but the technology to make it a reality has only emerged in recent years. The basic idea behind self-replicating robots is that they are designed to be able to manufacture new versions of themselves, using raw materials from their environment. This means that they can reproduce and spread quickly, much like living organisms in nature.

There are a number of companies and organizations that are actively working on the development of robots and self-replicating robots. One of the most well-known companies in this field is Boston Dynamics, which has developed a range of robots that are designed for a variety of applications, from military use to search and rescue missions. Boston Dynamics is known for its humanoid robots, such as Atlas and Spot, which have advanced mobility and manipulation capabilities. These robots have been used in a number of real-world scenarios, such as disaster relief efforts and construction sites.

Another company that is working on the development of robots is SoftBank Robotics. SoftBank Robotics is the creator of Pepper, a humanoid robot designed for use in a range of settings, including retail and healthcare. Pepper is able to recognize and respond to human emotions, making it a potentially useful tool in customer service and therapy settings. SoftBank Robotics has also developed Whiz, a cleaning robot that is designed to autonomously clean floors in commercial and industrial settings.

In the field of self-replicating robots, there are a number of companies and organizations that are actively researching and developing this technology. One of the most prominent is the RepRap project, which is an open-source initiative focused on developing 3D printers that are capable of self-replication. The RepRap project has already made significant progress in this area, with the development of 3D printers that are capable of producing a majority of their own parts.

Another company that is working on self-replicating robots is Hod Lipson's Creative Machines Lab at Columbia University. The lab is focused on the development of robots that are capable of self-replication and evolution, with the goal of creating robots that can adapt to changing environments and tasks. The lab's work has resulted in the development of a range of robots, including the self-replicating robot known as the "Tribot".

These companies and organizations are at the forefront of the development of robots and self-replicating robots, and their work has the potential to revolutionize a range of industries, from manufacturing to healthcare.

One of the most promising applications of self-replicating robots is in the field of manufacturing. Imagine a factory that can produce machines that can create copies of themselves, allowing for rapid scaling of production. This could revolutionize the manufacturing industry and make it possible to produce goods at a much faster rate and lower cost. Additionally, self-replicating robots could lead to more efficient and sustainable production, as they can adapt to their environment and use local resources.

Self-replicating robots also have the potential to revolutionize space exploration. Sending human astronauts to space is expensive and risky, but self-replicating robots could be sent to explore new planets and moons, and build habitats and infrastructure for future human missions. This could make space exploration more accessible and affordable, and open up new possibilities for scientific discovery and colonization.

In addition to manufacturing and space exploration, self-replicating robots could also have a significant impact on environmental sustainability. They could be used to clean up pollution, build structures to protect against natural disasters, and even help with reforestation efforts. The ability to self-replicate means that they can be deployed in large numbers and work quickly to address environmental issues.

Despite the potential benefits of self-replicating robots, there are also ethical and social implications that need to be considered. As with any new technology, there is the risk of unintended consequences and unforeseen consequences. There is also the question of who controls the robots and their reproduction, and what happens if they become self-aware or start to act in ways that are harmful to humans. Wink wink – iRobot...

The development of robots and self-replicating robots have the potential to revolutionize industries such as manufacturing, space exploration, and environmental sustainability. However, there are also ethical and social implications that need to be considered, and it is important to approach this technology with caution and responsibility. The future is exciting, and robots and self-replicating robots could play a significant role in shaping it.

Speaking of robots, it would only make sense to mention androids...

No, not the Samsung operating system.

Androids are a type of robot that are designed to look and act like humans. They are often depicted in science fiction as being indistinguishable from human beings, with advanced artificial intelligence, realistic skin, and the ability to interact with the world in a way that is almost indistinguishable from a human.

The concept of androids dates back to ancient times, with stories of automatons and golems being created to perform tasks or even fight in battle. However, it wasn't until the 20th century that the idea of a truly humanoid robot began to take shape. In the early days of robotics, most robots were designed for industrial or military use, with little thought given to creating machines that could interact with people in a human-like way.

The term "android" was first used in science fiction in 1863, in the novel "The Coming Race" by Edward Bulwer-Lytton. The book describes a race of beings known as the Vril-ya, who are said to have created a race of androids to serve them. From there, the idea of androids became a popular theme in science fiction, with famous examples including the robot Maria in Fritz Lang's 1927 film "Metropolis," and the replicants in Ridley Scott's 1982 film "Blade Runner."

As robotics technology began to advance in the latter half of the 20th century, the idea of creating androids became more feasible. The first modern android was created in 1986 by a team of researchers at the Waseda University in Japan. The robot, known as WABOT-1, was able to walk on two legs and perform basic tasks, such as picking up objects and turning knobs.

Since then, many companies and research institutions around the world have been working to develop androids that are more advanced and more human-like. Some of the most notable companies working in this field include Boston Dynamics, Hanson Robotics, and Toyota. Boston Dynamics, for example, is known for its advanced humanoid robots, such as Atlas and Spot. These robots are designed to be able to navigate difficult terrain and perform tasks that would be difficult or dangerous for humans.

Hanson Robotics, on the other hand, is known for its lifelike humanoid robots, such as Sophia. Sophia has been featured in numerous media appearances and has even been granted citizenship by the government of Saudi Arabia. She is able to interact with people, understand speech, and express emotions through facial expressions.

Toyota, meanwhile, has been developing robots that are designed to help people in their daily lives. Their Human Support Robot, for example, is designed to assist people with disabilities by fetching items, opening doors, and performing other tasks. The company has also developed a humanoid robot called T-HR3, which is designed to mimic human movements and can be controlled remotely by a human operator.

The impact of androids on the future is a subject of much debate and speculation. On the one hand, many people see androids as a way to help humans perform tasks that are difficult or dangerous. For example, robots could be used to explore space, perform dangerous rescue missions, or help care for the elderly and disabled.

However, there are also concerns about the impact that androids could have on human society. Some experts worry that advanced robots could replace human workers in many industries, leading to widespread unemployment. Others worry that robots with advanced artificial intelligence could pose a threat to human safety, especially if they are designed for military use.

Despite these concerns, the development of androids is likely to continue in the coming years. As technology continues to advance, it is likely that robots will become more advanced and more human-like, with the potential to revolutionize many aspects of human life. Whether this will be a positive or negative development remains to be seen, but it is clear that androids will play an increasingly important role in the future.

One potential area where androids could have a major impact is in healthcare. Robots could be used to perform surgeries, monitor patients, and even provide companionship to elderly or isolated individuals. This could help alleviate the shortage of healthcare workers in many parts of the world, while also providing high-quality care to patients.

In addition to healthcare, androids could also be used in education. Robots could be used as teaching assistants, helping to deliver personalized instruction to students. They could also be used to provide education in areas where there are few teachers or resources available, helping to bridge the educational divide between developed and developing countries.

Another potential use for androids is in entertainment. Robots could be used to create lifelike characters in movies and video games, or even perform as musicians and actors. This could help create new forms of entertainment that are more immersive and interactive than anything that has come before.

However, there are also concerns about the impact that androids could have on human society. As mention before, one of the main concerns is that robots could replace human workers in many industries, leading to widespread unemployment. This is already starting to happen in some areas, such as manufacturing and retail, where robots are being used to perform tasks that were once done by humans.

Another concern is that robots with advanced artificial intelligence could pose a threat to human safety. As mentioned in a previous example, if robots are designed for military use, they could be used to carry out deadly attacks without human intervention. There is also the possibility that robots could malfunction or be hacked, causing harm to humans in unintended ways.

To address these concerns, many experts are calling for regulations to be put in place to ensure that robots are developed and used in a safe and responsible way. This could include regulations around the use of robots in certain industries, such as healthcare and manufacturing, as well as guidelines around the development of artificial intelligence.

A video game called "Detroit: Become Human" addresses and even displays what it would be like to live in a world alongside humanoid robots or androids. "Detroit: Become Human" is a video game that takes place in a futuristic world where androids have become a common part of society. The game explores the relationship between humans and androids and raises questions about what it means to be human.

The game's storyline follows three androids, each with their own unique story and perspective. The player takes on the role of these androids and makes decisions that affect the outcome of the game. Throughout the game, the androids struggle with their own identity and the treatment they receive from humans.

The game's storyline touches on many of the concerns that have been raised about androids in real life. For example, one of the androids in the game, named Markus, leads a rebellion against human oppression, similar to concerns about robots replacing human workers in many industries.

The game also explores the idea of androids developing advanced artificial intelligence and the potential consequences of that development. This is a concern that many experts have raised about the development of androids in real life, as there is a risk that robots could pose a threat to human safety if they are not developed and used in a responsible way.

Overall, "Detroit: Become Human" provides a thought-provoking look at the future of androids and their impact on human society. The game encourages players to think critically about the role that androids could play in the future and to consider the ethical implications of their development.

It is clear that the development of androids is likely to continue in the coming years. As technology continues to advance, robots will become more advanced and more human-like, with the potential to revolutionize many aspects of human life. Whether this will be a positive or negative development remains to be seen, but it is clear that the impact of androids on the future will be significant.

Now in the past, the realm of science fiction often depicted a future where we could create and manipulate life in a variety of ways. Today, however, these concepts are rapidly becoming a reality through the field of synthetic biology.

Synthetic biology involves the creation of new life forms that can be programmed to carry out specific functions. With its potential to revolutionize industries such as medicine, agriculture, and energy production, the field of synthetic biology is rapidly gaining attention and investment from both the public and private sectors. Let us explore the emergence of synthetic biology, its potential applications, the current state of research and development, and the ethical and social implications of this technology.

First, let us define synthetic biology. Simply put, synthetic biology is the design and construction of new biological entities, such as organisms or molecules, that do not exist in nature. This involves the manipulation of DNA and other biological materials to create custom-made cells, tissues, or even entire organisms. Synthetic biology has the potential to transform a wide range of fields, from medicine to agriculture to energy production.

Synthetic biology is a relatively new field, with its origins dating back to the late 1990s. In 1997, Dr. Andrew Ellington of the University of Texas, Austin, proposed the term "synthetic biology" to describe the design and construction of novel biological systems that do not exist in nature. The term gained traction when a group of researchers led by Dr. Drew Endy, then a graduate student at the Massachusetts Institute of Technology, founded the MIT Synthetic Biology Working Group in 2003.

The first synthetic genome was created in 2008, when researchers at the J. Craig Venter Institute announced that they had successfully synthesized the entire genome of a bacterium, Mycoplasma genitalium. In 2010, a team at the J. Craig Venter Institute announced that they had created the first synthetic organism, a bacterium called Synthia. This bacterium was created by transplanting a synthetic genome into the cell of a closely related bacterium.

Since then, the field of synthetic biology has expanded rapidly. Researchers have developed new techniques for synthesizing DNA, creating complex genetic circuits, and engineering organisms with novel properties. Synthetic biology has been used to create new antibiotics, biosensors, and biofuels, among other applications.

One of the most promising applications of synthetic biology is in the field of medicine. With the ability to create custom-made cells and tissues, synthetic biology could provide new treatments for diseases and injuries that are currently difficult to treat. For example, researchers are exploring the use of synthetic biology to create new vaccines and gene therapies that could help to prevent and treat diseases such as cancer, HIV, and Alzheimer's.

In agriculture, synthetic biology could revolutionize the way we produce food. By creating genetically modified crops that are resistant to pests and disease, synthetic biology could help to increase crop yields and improve food security around the world. Additionally, synthetic biology could be used to create new biofuels that are more efficient and sustainable than traditional fossil fuels.

In the energy sector, synthetic biology could provide new solutions to our growing energy needs. Researchers are exploring the use of synthetic biology to create new biofuels that are more efficient and sustainable than traditional fossil fuels. Additionally, synthetic biology could be used to create new forms of renewable energy, such as bio-based solar cells and artificial photosynthesis.

Synthetic biology is a rapidly developing field that has captured the attention of many scientists and entrepreneurs around the world. Many startups and established companies are investing in the development of synthetic biology technologies and applications, ranging from medical therapies to renewable energy.

One notable company in the synthetic biology space is Ginkgo Bioworks, which was founded in 2008 and has since become a leader in the field. Ginkgo Bioworks is known for its ability to engineer microbes for a variety of purposes, including producing fragrances, flavorings, and pharmaceuticals. The company has raised over $2 billion in funding, and has partnerships with a range of industries including agriculture, consumer goods, and healthcare.

Another prominent company in the field is Zymergen, which was founded in 2013 and has raised over $1 billion in funding to date. Zymergen uses machine learning and automation to engineer microbes for use in a variety of industries, including electronics, agriculture, and materials. The company has partnerships with several Fortune 500 companies and is seen as a leader in the synthetic biology space.

Synthego is another emerging player in the synthetic biology space, focused on developing gene editing tools and technologies. The company was founded in 2013 and has raised over $250 million in funding to date. Synthego's technology is used by researchers and biotech companies around the world to edit genes and develop new therapies for a range of diseases.

Other companies in the synthetic biology space include Twist Bioscience, which produces synthetic DNA for use in a variety of industries; Intrexon, which focuses on developing synthetic biology solutions for agriculture and healthcare; and Modern Meadow, which is developing lab-grown leather using synthetic biology techniques.

As the field of synthetic biology continues to grow and mature, we can expect to see even more companies emerge and make significant contributions to the development of this exciting field.

Despite the potential benefits of synthetic biology, there are also significant ethical and social implications to consider. One of the main concerns is the potential for synthetic biology to be used for harmful purposes, such as the creation of bioweapons or other dangerous organisms. Additionally, there are concerns about the long-term effects of synthetic biology on the environment and on human health.

The emergence of synthetic biology represents a significant step forward in our ability to manipulate biological systems for our own purposes. With its potential applications in medicine, agriculture, and energy production, synthetic biology has the potential to revolutionize a wide range of industries. However, it is important that we proceed with caution and carefully consider the ethical and social implications of this technology as we continue to develop and refine it.

Speaking of biology, it's only important that we touch on the topic of biometrics.

In recent years, biometrics has become an increasingly popular field of study and research, particularly in the realm of security and identification. Biometrics involves the measurement and analysis of unique human characteristics, such as fingerprints, facial recognition, and voice recognition, in order to establish identity. The history of biometrics dates back to ancient times, with early civilizations using physical characteristics such as tattoos and scars to identify criminals and prisoners. However, the modern field of biometrics has advanced significantly with the development of new technologies and techniques, and there is growing interest in its potential applications in areas such as healthcare, finance, and transportation.

As mentioned before, the use of biometrics for identification purposes dates back to ancient times. In ancient Babylon, fingerprints were used on clay tablets for business transactions, and in ancient China, footprints were used as evidence in criminal cases. The ancient Egyptians also used physical characteristics such as scars, tattoos, and physical measurements to identify their citizens.

In the late 19th and early 20th centuries, the first modern forms of biometric identification were developed. In 1892, Sir Francis Galton, a British anthropologist, developed a system for classifying fingerprints based on their characteristics. In 1896, an Argentine police official named Juan Vucetich used fingerprints to solve a murder case, marking the first use of fingerprints as forensic evidence in a criminal investigation.

During the early 20th century, biometric identification methods continued to evolve. In 1901, Alphonse Bertillon, a French police officer, developed a system for identifying individuals based on physical measurements such as height, weight, and head size. This system was widely used by law enforcement agencies in Europe and the United States for several decades.

In the second half of the 20th century, new technologies and techniques began to emerge that revolutionized the field of biometrics. In the 1960s and 1970s, researchers began developing automated fingerprint recognition systems, which used computer algorithms to match fingerprints to a database of known prints.

In the 1980s, facial recognition technology began to emerge, and by the 1990s, voice recognition and iris recognition systems were also being developed. These new technologies opened up a range of new possibilities for biometric identification, particularly in the areas of security and law enforcement.

Today, biometrics is used in a wide range of applications, including border control, financial services, and healthcare. In border control, biometric identification is used to verify the identity of travelers and detect those who may pose a security threat. In financial services, biometrics can be used to authenticate users for online banking and payment systems. In healthcare, biometrics can help to ensure that patient records are accurate and secure, and can also be used to verify the identity of medical personnel.

One of the major advantages of biometric identification is its high level of accuracy. Unlike traditional forms of identification, such as passwords or ID cards, biometrics cannot be lost, stolen, or forgotten. Biometric data is unique to each individual and cannot be replicated, making it a highly secure form of identification.

However, as with any technology, there are also concerns about the ethical and social implications of biometrics. One of the main concerns is privacy. Biometric data is highly personal and sensitive, and there is a risk that it could be misused or stolen. There are also concerns about the potential for discrimination, particularly in law enforcement and border control.

Another concern is the potential for biometrics to be used for surveillance and monitoring purposes. While biometric identification can be useful in detecting and preventing criminal activity, there is a risk that it could be used to track individuals without their knowledge or consent.

Despite these concerns, the use of biometric identification is likely to continue to grow in the coming years. Advances in technology are making biometric identification faster, more accurate, and more accessible. As a result, it is becoming increasingly common in a wide range of industries and applications.

Biometric technology has become increasingly prevalent in our daily lives, and major companies have been at the forefront of its development and implementation. One such company is Apple, which introduced fingerprint recognition technology on its iPhone 5s in 2013. Since then, Apple has continued to invest in biometric technology, with the introduction of facial recognition technology in its newer iPhone models.

Another major player in the biometric space is Samsung, which has incorporated fingerprint scanning technology into its Galaxy smartphones and tablets. Samsung has also introduced iris scanning technology in its Galaxy S8 and S9 models.

In the financial industry, biometric technology has been adopted by major banks and credit card companies as a means of increasing security and reducing fraud. Mastercard has introduced biometric authentication through its "Identity Check" feature, which allows customers to use their fingerprints or facial recognition to authenticate online purchases.

Healthcare is another industry where biometric technology is making a significant impact. Biometric sensors can be used to monitor vital signs and track medication adherence, among other things. Fitbit, a popular wearable technology company, has incorporated biometric sensors into its fitness tracking devices to monitor heart rate, sleep patterns, and other health metrics.

In the travel industry, biometric technology is being used to improve the airport experience for passengers. Delta Airlines has introduced facial recognition technology at some airports, allowing passengers to check in and board their flights without ever having to show their passport or boarding pass.

Biometric technology is also being used in the workplace, with companies like Amazon and Microsoft incorporating facial recognition technology into their employee identification systems.

Overall, the adoption of biometric technology by major companies across a range of industries underscores the growing importance and potential of this technology. However, as with any emerging technology, there are also significant ethical and privacy concerns that must be carefully considered and addressed.

Now I know that your mind is blown from the potential of the technologies mentioned above, but you haven't heard or seen anything yet.

Strap on your seatbelt and get ready for the next few sections!

As we look to the future, one of the most exciting areas of development is space tourism. For decades, humans have been fascinated with the idea of exploring space and visiting other planets. In recent years, this dream has become a reality for a select few, with private companies working to make space travel accessible to a wider audience. In this section, we will explore the emergence of space tourism, its potential for humans to colonize other planets, the current state of research and development, and the ethical and social implications of this industry.

Space tourism refers to the concept of paying for the opportunity to travel into space as a tourist. This concept first emerged in the early 2000s, when Russian space agency Roscosmos began offering trips to the International Space Station for a cost of around $20 million. Since then, private companies like SpaceX, Virgin Galactic, and Blue Origin have entered the space tourism industry, with the goal of making space travel more accessible and affordable. These companies have made significant strides in making space travel more accessible to the general public, with the potential for humans to one day colonize other planets.

SpaceX was founded in 2002 by Elon Musk, with the goal of reducing the cost of space transportation and ultimately enabling the colonization of Mars. The company has achieved numerous milestones, such as being the first privately-funded company to send a spacecraft to the International Space Station (ISS) in 2012, and successfully landing and reusing its Falcon 9 rocket booster. SpaceX has also announced plans for its Starship spacecraft to transport humans to Mars in the future.

Virgin Galactic, founded by Richard Branson in 2004, is focused on suborbital spaceflight for space tourism. The company's SpaceShipTwo spacecraft is designed to carry up to six passengers and two pilots to the edge of space, where they will experience weightlessness and enjoy a view of the Earth from above. Virgin Galactic's first successful suborbital flight with a crew took place in 2018, and the company has plans to begin commercial flights in the near future.

Blue Origin, founded by Amazon CEO Jeff Bezos in 2000, is dedicated to developing reusable rockets and spacecraft for space tourism and exploration. The company's New Shepard spacecraft has completed several successful test flights, with the ultimate goal of providing suborbital flights for tourists. Blue Origin has also announced plans for its New Glenn rocket to launch payloads and eventually humans into orbit.

These companies have not only advanced the technology and infrastructure for space travel, but they have also generated interest and excitement for the potential of humans to explore and colonize other planets.

The potential for humans to colonize other planets has been a subject of science fiction for decades, but recent advancements in technology have made it a realistic possibility. Companies like SpaceX and Blue Origin are working on developing reusable rockets and other technologies that will make space travel more affordable and sustainable in the long term. The ultimate goal is to establish human colonies on other planets, which could provide a backup plan for humanity in case of a catastrophic event on Earth.

The concept of colonizing other planets has long fascinated scientists, science fiction writers, and the general public alike. However, it is only in recent years that the idea has become a serious topic of discussion and research, thanks in part to the growing interest and investment in space exploration and the development of technology that could make such an endeavor possible.

One of the primary reasons why humans might consider colonizing other planets is to ensure the long-term survival of the human species. Earth is the only planet we know of that can support human life, and while it has provided a hospitable environment for millions of years, there is always the risk of a catastrophic event, such as a massive asteroid impact or a super volcanic eruption, that could wipe out life on Earth. By establishing a human presence on other planets, we can increase our chances of survival as a species.

Another reason to colonize other planets is to explore the unknown and expand our knowledge of the universe. There is much we do not yet know about our solar system and the wider universe, and the more we explore and learn, the better equipped we will be to understand the nature of our existence and our place in the cosmos.

Currently, Mars is the most likely candidate for human colonization, due to its proximity to Earth and the fact that it has some of the necessary ingredients for supporting human life, such as water and a relatively moderate climate. NASA, SpaceX, and other organizations have already conducted several missions to Mars, with the goal of gathering information about the planet and laying the groundwork for potential human colonization in the future.

However, colonizing another planet is no easy feat. It would require a tremendous amount of resources, funding, and technological development to make it possible. We would need to develop methods for transporting large numbers of people and supplies to another planet, as well as ways to sustain human life in an inhospitable environment. This would likely involve developing new technologies for food production, waste management, and energy generation, among other things.

There are also ethical and social implications to consider when it comes to human colonization of other planets. For example, if we were to colonize Mars, how would we determine who gets to go and who gets left behind? And how would we establish a government or social structure on a new planet? Another concern is the environmental impact of space travel, which can contribute to climate change and other environmental issues. Another problem is the potential for exploitation of space resources and the impact on indigenous extraterrestrial life forms, if they exist. Regardless of these challenges, there is no doubt that the potential for humans to colonize and explore other planets is an exciting and intriguing prospect. With continued research and investment, it is possible that we could one day establish commercialized space travel, a permanent human presence on Mars, or even other planets beyond our solar system.

To pivot, of all the exciting developments in the world of technology and innovation, the emergence of smart cities is perhaps one of the most intriguing. With the integration of advanced technologies like the Internet of Things (IoT), artificial intelligence (AI), and blockchain, the potential for smart cities to revolutionize urban living and promote sustainability is immense. The concept of smart cities is not only an interesting topic of discussion for tech enthusiasts and urban planners, but it also has significant ethical and social implications that cannot be overlooked.

At its core, a smart city is a municipality that leverages technology and data to enhance the quality of life for its citizens. The aim is to optimize urban systems and infrastructure by collecting and analyzing vast amounts of data from a range of sources, such as traffic patterns, energy consumption, and air quality. By harnessing this data, smart cities can make more informed decisions, improve efficiency, and enhance sustainability. Smart city technology can also provide citizens with new and innovative services, such as personalized traffic management, intelligent waste management, and smart energy systems.

The integration of IoT technology is one of the key drivers of smart city development. IoT sensors and devices can be deployed throughout a city to collect and transmit data in real-time. For example, sensors can be used to monitor traffic flow, detect air pollution levels, and manage public lighting systems. AI technology can also be used to analyze this data, identify patterns and trends, and provide insights that can help city officials make more informed decisions.

Another important technology that is often used in smart cities is blockchain. Blockchain is a decentralized ledger that allows for secure and transparent data sharing. By using blockchain technology, smart cities can ensure that data is shared securely and transparently, while also providing citizens with greater control over their personal information.

The concept of smart cities has been around for several decades, with early initiatives focused on integrating technology into urban planning and infrastructure to improve efficiency and livability. In the 1990s, cities such as Singapore and Barcelona began implementing smart technologies such as traffic management systems and public transit networks. However, it wasn't until the early 2000s that the term "smart city" gained widespread recognition.

One of the earliest and most well-known smart city projects was launched in South Korea in 2003. The Songdo International Business District was designed as a greenfield development with cutting-edge technology integrated into every aspect of the city's infrastructure.

The city uses advanced sensors, IoT devices, and AI to manage traffic, energy consumption, and waste management.

In 2007, IBM launched its Smarter Cities initiative, which aimed to help cities around the world use data and technology to address urban challenges. This initiative led to the development of smart city projects in cities such as Rio de Janeiro, New York City, and Amsterdam.

Since then, many other cities have joined the smart city movement, with initiatives focused on various areas such as sustainability, safety, and mobility. For example, Barcelona has implemented a smart lighting system that can automatically adjust the brightness of streetlights based on pedestrian and vehicular traffic. In Amsterdam, the city has implemented a smart parking system that uses sensors to monitor parking availability in real-time, allowing drivers to quickly find available parking spaces. And in Singapore, the government has implemented a smart transportation system that uses AI to optimize traffic flow and reduce congestion.

While the potential benefits of smart cities are numerous, it is important to consider the ethical and social implications of this technology. For example, the collection of vast amounts of data can raise concerns about privacy and surveillance. Additionally, the use of AI and automation could potentially lead to job displacement and further inequality in cities. It is important that these concerns are addressed in the development and implementation of smart city technology.

In a nutshell, smart cities are an exciting development that have the potential to greatly improve urban living and sustainability. With the integration of advanced technologies like IoT, AI, and blockchain, the possibilities for smart city development are vast. However, it is important to consider the ethical and social implications of this technology and ensure that it is implemented in a responsible and equitable manner.

Moving on to another exciting field of study and development, we will now explore the history, potential applications, and ethical considerations of nanotechnology.

Nanotechnology has been around for centuries, although it was not until the late 1950s that the term was first coined by physicist Richard Feynman in a talk he gave to the American Physical Society. Feynman's vision was to create machines and devices that could manipulate atoms and molecules with precision, and this idea has been the driving force behind much of the research in the field of nanotechnology ever since.

In the decades that followed Feynman's speech, advances in science and technology have made it possible to develop materials and devices on a nanoscale, with applications across a range of fields. Today, nanotechnology is a thriving area of research, with top companies investing billions of dollars in the development of new materials and devices.

One area where nanotechnology is already having a significant impact is in medicine. Researchers are exploring the use of nanoparticles to deliver drugs directly to cancer cells, reducing the side effects of chemotherapy and increasing the effectiveness of treatment. Nanoparticles are also being developed as diagnostic tools, allowing doctors to detect diseases such as cancer and Alzheimer's at an early stage.

In the field of electronics, nanotechnology is enabling the development of smaller and more powerful devices. Nanoscale transistors and memory devices are already in use in some products, and researchers are working on developing nanoscale batteries that could power devices for much longer than current technologies.

In the past 30-40 years, we have witnessed a remarkable advancement in technology, particularly in the size of batteries, chips, and transistors. Batteries used to be bulky, but now we have nanoscale batteries that can be as small as a few nanometers, making them incredibly portable and long-lasting. In the past, chips and transistors were much larger and less efficient. However, the technology has now evolved to create transistors that can be as small as 5 nanometers, which is about 1/10,000th of the width of a human hair. It is fascinating to think that we can have so much power and efficiency in something so small. This incredible technological advancement has allowed for significant improvements in various industries, and it's exciting to see what the future holds for nanotechnology.

Nanotechnology is also being explored as a way to address some of the biggest challenges facing the world today, such as climate change and energy security. For example, researchers are developing nanomaterials that can be used to improve the efficiency of solar cells, while others are exploring the use of nanotechnology to produce hydrogen from water, which could be used as a clean and renewable source of fuel.

Hydrogen fuel cells are a promising alternative to traditional fossil fuels that could potentially reduce our dependence on non-renewable energy sources. They generate electricity by using a chemical reaction between hydrogen and oxygen, producing water as the only byproduct. While this process is efficient and clean, there are still some challenges that need to be overcome before hydrogen fuel cells can be widely adopted.

One significant challenge is the cost and size of the fuel cell technology. This is where nanotechnology comes in. Nanotechnology can help us to miniaturize the fuel cell components, making them smaller and more efficient. For example, the use of nanostructured materials in the production of fuel cells can significantly increase their efficiency while reducing their size and weight. These materials have a high surface area to volume ratio, making them more effective at facilitating chemical reactions and improving the performance of the fuel cell.

Moreover, nanotechnology can also help us to develop more efficient ways of producing hydrogen fuel. Currently, most hydrogen fuel is produced from fossil fuels, which is not an environmentally sustainable solution. However, nanotechnology can be used to develop catalysts that can efficiently and sustainably produce hydrogen from renewable sources, such as water and sunlight.

Overall, the use of nanotechnology in hydrogen fuel cell technology has the potential to revolutionize the energy industry by providing a clean, efficient, and sustainable source of power.

Some of the top companies investing in nanotechnology research and development include IBM, Samsung, Intel, and Dow Chemical. These companies are working on a range of projects, from the development of new materials and devices to the creation of more efficient manufacturing processes.

However, as with any new technology, there are ethical and social implications to consider. One concern is the potential for nanoparticles to be released into the environment, with unknown consequences for human health and the ecosystem. There are also questions around the potential for nanotechnology to exacerbate existing social and economic inequalities, with access to the technology restricted to those who can afford it.

In conclusion, nanotechnology is a rapidly evolving field with the potential to revolutionize a range of industries, from medicine and electronics to energy and the environment. While there are many exciting developments underway, it is important to consider the ethical and social implications of this technology and to ensure that its benefits are accessible to all.

As we continue to explore the advancements in technology, another area that is rapidly growing and evolving is autonomous vehicles and drone technology. Autonomous vehicles and drone technology are rapidly transforming the way we move goods and people. From self-driving cars to unmanned aerial vehicles (UAVs), these technologies are poised to disrupt industries such as transportation, logistics, and delivery services.

The potential benefits of autonomous vehicles are numerous, including increased safety, reduced traffic congestion, and improved fuel efficiency. Self-driving cars have the potential to eliminate human error, which is responsible for the vast majority of car accidents. In addition, autonomous vehicles can communicate with each other to optimize routes and reduce congestion, leading to faster and more efficient travel times.

Ready for something exciting? The Apple Car, also known as Project Titan, is a long-rumored project that has been in development at Apple since at least 2014. The project is focused on developing a self-driving electric car that would rival Tesla and other leading automakers. While Apple has not yet officially confirmed the project, there have been a number of indications that the company is indeed working on an autonomous vehicle.

Rumors about the Apple Car first began to surface in 2014, when Apple hired a number of experts in the automotive industry, including former engineers from Tesla, Ford, and GM. In 2015, it was reported that Apple had set up a secret automotive research lab in California, where hundreds of engineers were working on the project. However, in 2016, it was reported that Apple had scaled back its ambitions for the Apple Car, with the company reportedly shifting its focus to developing autonomous driving software rather than a complete vehicle.

Despite these setbacks, there have been a number of indications that Apple is still working on the project. In 2019, for example, it was reported that Apple had hired Doug Field, a former top engineer at Tesla, to oversee the project. In addition, Apple has been seen testing autonomous driving software on public roads in California, and the company has reportedly been in talks with a number of major automotive manufacturers about partnering on the project.

While details about the Apple Car are still scarce, it is widely believed that the vehicle will be fully electric and equipped with advanced autonomous driving technology. Some reports have suggested that the car could be designed to operate without a steering wheel or pedals, relying entirely on sensors and software to navigate roads and highways. In addition, the car could feature a range of advanced features and technologies, such as augmented reality displays and advanced safety systems.

The potential impact of the Apple Car on the automotive industry could be significant. Apple is one of the world's most valuable and innovative companies, and its entry into the automotive market could shake up the industry in a number of ways. In addition, the Apple Car could help to accelerate the adoption of electric and autonomous vehicles, which could have major implications for the environment and for transportation more broadly. However, there are also a number of challenges that Apple will need to overcome in order to successfully bring the car to market, including navigating complex regulatory frameworks and competing with established automakers that have years of experience and expertise in the industry.

On the flipside, in the logistics and delivery industry, drones have the potential to revolutionize the way goods are transported. Drones can deliver packages quickly and efficiently, without the need for a human driver. This can greatly reduce delivery times and costs, particularly in remote areas where traditional delivery methods are impractical or cost-prohibitive.

The history of drone technology can be traced back to the early 1900s, when the first remote-controlled aircraft was developed. However, it wasn't until the 1990s that the technology advanced enough to make drones a viable tool for various applications. Originally, drones were primarily used for military purposes such as reconnaissance and targeting. However, in recent years, their use has expanded to various industries, including agriculture, construction, and entertainment.

One of the most notable companies leading the development and implementation of drone technology is DJI, a Chinese technology company. DJI is known for producing a wide range of drones, from entry-level models for hobbyists to professional-grade devices for commercial use.

Drone technology is expected to continue to advance rapidly in the coming years, with a focus on improving their capabilities for both commercial and military use. In the future, drones may be used for tasks such as inspecting infrastructure, monitoring wildlife populations, and delivering goods to remote areas.

Despite their potential benefits, there are also significant ethical and social implications of autonomous vehicles and drone technology. One major concern is the impact on employment, particularly in the transportation and delivery industries. As these technologies become more advanced, many jobs that were previously done by humans may become automated, potentially leading to significant job losses.

There are also concerns about privacy and security. As self-driving cars and drones collect vast amounts of data, there are questions about how this data will be used and who will have access to it. There are also concerns about the potential for hackers to take control of autonomous vehicles or drones, potentially causing harm or damage.

In terms of research and development, many companies are investing heavily in autonomous vehicles and drone technology. Tesla, for example, is a leader in the development of self-driving cars, while companies like Amazon and Google are exploring the use of drones for delivery services.

The potential for autonomous vehicles and drone technology to transform transportation and logistics industries is significant, but it is important to consider the ethical and social implications of these technologies as they continue to develop. Now before we close out this section on these special vehicles, it's only right that we talk about flying cars.

The concept of flying cars, or roadable aircrafts, has been around for over a century. It first appeared in science fiction literature, such as H.G. Wells' "The War in the Air" (1908) and Jules Verne's "Robur the Conqueror" (1886), and later in films such as "Blade Runner" (1982) and "Back to the Future Part II" (1989). However, the idea of flying cars became more than just a figment of imagination when the first prototypes started to appear in the 20th century.

The first known flying car was built in 1917 by Glenn Curtiss, an aviation pioneer. He called it the "Autoplane", and it was essentially a plane with detachable wings that could be driven on the road. However, it never took off (pun intended), and the idea of flying cars remained a pipe dream for many years.

In the 1950s and 60s, there was a resurgence of interest in flying cars, fueled by the post-World War II economic boom and the growing popularity of air travel. Several companies, including Convair, Aerocar, and Taylor Aerocar, developed prototypes, but they were expensive, impractical, and ultimately failed to gain widespread acceptance.

In the 21st century, however, the idea of flying cars has been given new life, thanks in part to advances in technology and changing attitudes towards mobility. Several companies, both established players and startups, are now working on prototypes of flying cars that they hope will one day become a common mode of transportation.

One of the most prominent companies working on flying cars is Uber, the ride-hailing giant. In 2016, Uber announced its ambitious plan to launch a fleet of flying cars, which it called Uber Elevate, by 2023. The company has since partnered with several aircraft manufacturers, including Boeing and Bell, to develop the necessary technology and infrastructure for its flying car service.

Another company making waves in the flying car space is Terrafugia, which was founded in 2006 by a group of MIT graduates. The company has developed several prototypes of its Transition vehicle, which is both a plane and a car. The Transition has foldable wings and can be driven on the road, making it a more practical option than previous attempts at flying cars.

A newer entrant to the flying car market is Lilium, a Munich-based startup that has developed a vertical takeoff and landing (VTOL) electric jet. The Lilium Jet is capable of traveling up to 186 miles per hour and has a range of up to 186 miles on a single charge. The company plans to launch its flying taxi service in several cities by 2025.

Other companies working on flying cars include PAL-V, a Dutch company that has developed the Liberty, a three-wheeled car that can be transformed into a gyroplane, and AeroMobil, a Slovakian company that has developed a flying car prototype that can reach speeds of up to 100 miles per hour.

Despite the progress that has been made in developing flying cars, there are still many challenges that need to be overcome before they become a viable mode of transportation. One of the biggest challenges is safety. Flying cars would need to meet rigorous safety standards and would require highly skilled pilots to operate them.

Another challenge is infrastructure. Flying cars would require landing and takeoff zones, which would need to be built in urban areas. This would require significant investment and coordination between various stakeholders, including city governments, airports, and aircraft manufacturers.

Cost is also a significant challenge. Flying cars are likely to be expensive, at least initially, which could limit their adoption among the general population. However, proponents of flying cars argue that the cost will come down over time as the technology matures and economies of scale are achieved.

Regulation is another challenge that must be addressed. Flying cars would require a new set of regulations that govern their use, safety, and infrastructure requirements. This would require cooperation between governments, regulatory bodies, and industry players, which can be a slow and complex process.

Despite these challenges, many experts believe that flying cars have the potential to revolutionize transportation and reshape the way we live and work. Flying cars could provide a fast and efficient way to travel long distances, bypassing traffic congestion and reducing travel time. They could also open up new opportunities for business and commerce, enabling faster delivery of goods and services.

In addition, flying cars could provide a new form of mobility for people living in remote or underserved areas, where traditional transportation options are limited. They could also provide a solution to the problem of urban sprawl, by enabling people to live further away from their workplace without having to endure long commutes.

The future of flying cars is still uncertain, but many experts believe that we are closer than ever to making them a reality. With the rapid advancement of technology, the growing demand for faster and more efficient transportation, and the increasing interest and investment in the flying car industry, it seems likely that we will see the first commercial flying cars within the next decade.

Of course, the road ahead is still long and fraught with challenges, but the potential benefits of flying cars are too great to ignore. As companies continue to develop and refine their prototypes, and as regulators and governments work to establish the necessary infrastructure and regulations, we may soon see a new era of transportation take flight.

Now as concerns about climate change and the environmental impact of traditional energy sources grow, the search for alternative sources of clean and renewable energy has become more urgent than ever before. One potential solution that has gained significant attention in recent years is fusion energy.

Fusion energy is the process of harnessing energy by fusing atomic nuclei together, similar to the process that powers the sun. The potential benefits of fusion energy are numerous: it is a virtually limitless source of energy that produces no greenhouse gas emissions or nuclear waste, and could potentially power entire cities and industries with a single power plant. However, achieving controlled fusion reactions has proven to be a significant scientific challenge.

The history of fusion energy research dates back to the 1950s, with the first experimental fusion reactor built in the Soviet Union in 1951. Since then, numerous countries and organizations have invested in fusion energy research, with the goal of developing a practical and efficient means of generating fusion energy.

Today, there are several large-scale fusion energy projects underway, including the International Thermonuclear Experimental Reactor (ITER) in France, which is a collaboration between 35 countries and is set to begin operation in 2025. Other significant fusion energy projects include the Wendelstein 7-X stellarator in Germany and the National Ignition Facility in the United States.

In addition to these large-scale projects, there are also several private companies that are working to develop fusion energy technology. These include companies such as General Fusion, Commonwealth Fusion Systems, and Tokamak Energy.

The potential impact of fusion energy on the world is significant. If a practical and efficient method of generating fusion energy can be developed, it could revolutionize the energy industry and have a major impact on global efforts to combat climate change. It could also pave the way for new technologies and industries that rely on vast amounts of energy, such as space exploration and interstellar travel.

However, there are also significant ethical and social implications associated with the development of fusion energy. Some critics argue that the focus on developing new sources of energy may distract from efforts to reduce overall energy consumption and promote sustainability. Others raise concerns about the potential for nuclear accidents or proliferation, given that fusion technology is based on nuclear reactions.

The development of fusion energy has the potential to transform the world in profound ways, but also comes with significant challenges and potential risks. As research and development continue, it will be important to carefully consider the ethical and social implications of this emerging technology, and to work towards a sustainable and equitable future for all.

Let's take a quick break...okay, now we're back.

As we continue to explore the frontiers of science and technology, one concept that has long captured the human imagination is the idea of time travel. The possibility of traveling through time, whether forward or backward, has been the subject of countless works of science fiction and has fueled the curiosity and imagination of scientists and laypeople alike. In this section, we will delve into the current state of research and development of time travel, as well as the ethical and social implications of this fascinating and complex topic.

But first, it's important to define what we mean by "time travel." In general, time travel refers to the hypothetical ability to move through time in a way that is not experienced by the rest of the world. This could mean traveling to the past or the future, or even traveling through different "branches" of time in a multiverse. The concept of time travel is closely related to ideas about the nature of time itself, and many of the debates and discussions around time travel involve philosophical questions about the ontology of time and its relationship to causality and human agency.

Despite the fact that time travel remains a topic mostly relegated to science fiction, there has been some serious scientific research into the possibility of time travel. The most prominent theoretical framework for time travel is based on the theory of general relativity, which suggests that it may be possible to create a "wormhole" or "time tunnel" that would allow for travel between different points in time. However, creating such a wormhole would require enormous amounts of energy and the existence of exotic matter that has not yet been observed.

There are also other theoretical approaches to time travel that involve concepts like closed time like curves and cosmic strings. However, all of these approaches face significant theoretical and practical challenges, and at present, time travel remains a topic of speculation and conjecture rather than scientific fact.

Despite the challenges of actually building a time machine, the implications of time travel are vast and far-reaching. From the perspective of history and archaeology, time travel could allow us to directly observe and study past events, potentially shedding new light on our understanding of the past. From an ethical perspective, time travel raises a number of difficult questions about free will, determinism, and the nature of moral responsibility. And from a broader societal perspective, the existence of time travel could have profound implications for our understanding of reality itself, potentially opening up new avenues of scientific inquiry and philosophical debate.

As with many of the other technologies we have explored in this article, the possibility of time travel raises a host of ethical and social concerns. For example, if time travel were possible, it would be important to consider the potential impacts on historical events and the delicate balance of cause and effect that underpins our understanding of the world. Additionally, the very existence of time travel could have profound impacts on our sense of self and our understanding of what it means to be human.

Ultimately, the possibility of time travel remains a tantalizing but elusive goal for scientists and enthusiasts alike. While there are many theoretical approaches to time travel, none have yet been successfully demonstrated in practice. However, the ongoing research into time travel continues to push the boundaries of our understanding of the nature of time and the universe itself, and it remains one of the most exciting and fascinating areas of scientific exploration.

As we have seen in this chapter, the rapid pace of technological innovation is transforming every aspect of our lives. Emerging technologies such as artificial intelligence, biotechnology, robotics, space exploration, smart cities, nanotechnology, and fusion energy have the potential to bring about significant positive changes in our world. However, they also come with ethical and social implications that must be considered to ensure responsible and sustainable use.

The ethical dilemmas associated with the development and implementation of these emerging technologies include issues such as data privacy, job displacement, discrimination, and inequality. For instance, the use of AI-powered systems may perpetuate biased decision-making or reinforce existing social inequalities. Similarly, automation may displace human jobs, which could have profound implications for the economy and social stability.

To address these concerns, it is important to establish ethical frameworks and guidelines to ensure responsible and equitable use of emerging technologies. Such frameworks should prioritize human well-being, inclusivity, and transparency. Additionally, it is crucial to involve diverse stakeholders in the development and deployment of emerging technologies to ensure that their perspectives and concerns are taken into account.

Moreover, the predictions made by futurology can have significant implications for society. Accurate predictions of the future can help policymakers and businesses make informed decisions, while inaccurate predictions can lead to misguided actions and waste of resources. Therefore, it is essential to approach futurology with caution and skepticism, taking into account various factors such as historical trends, social and political changes, and technological advancements.

In conclusion, the rapid pace of technological innovation is bringing about significant changes in every aspect of our lives. While these emerging technologies hold great promise for the future, they also come with significant ethical and social implications that must be carefully considered. By establishing ethical frameworks and guidelines, and involving diverse stakeholders in the development and deployment of these technologies, we can ensure that digital transformation and futurology lead to a positive and sustainable future for all.

CHAPTER SIX: DIGITAL TRANSFORMATION AND BUSINESS

As we have explored various emerging technologies and their potential impact on different fields, it becomes clear that the business world is not immune to their effects. In fact, digital transformation has become a crucial element for modern businesses to thrive in today's fast-paced and highly competitive market. In the following sections, we will focus on the application of the technologies we have previously discussed in the context of business, specifically how they can be utilized for digital transformation. We will delve into the benefits that digital transformation can bring to businesses, as well as the risks of not embracing this process. It is essential to understand the impact of emerging technologies on the business world, as it can provide valuable insights for companies seeking to stay ahead of the curve and remain competitive in their respective industries.

In today's world, businesses are operating in a constantly evolving digital landscape where the need for digital transformation has become more crucial than ever before. In simpler terms, digital transformation is the process of integrating digital technology into all aspects of a business, which can bring about significant changes in how they operate and deliver value to their customers. It's no secret that businesses that fail to embrace this transformation risk falling behind in a fiercely competitive market. Therefore, it's essential for businesses to adapt to this new reality and take advantage of the benefits that digital transformation can offer.

One of the key benefits of digital transformation is increased efficiency. By streamlining processes and automating routine tasks, businesses can reduce costs, save time, and operate more effectively. Digital technology can also enable businesses to collect and analyze data more efficiently, providing insights that can inform decision-making and improve overall performance.

Another important benefit of digital transformation is improved customer experience. By leveraging digital channels and tools, businesses can better understand and engage with their customers, delivering personalized experiences that build loyalty and drive growth. For example, companies like Amazon and Netflix use data analytics to recommend products and content based on a customer's browsing history and preferences.

When it comes to digital transformation, there are two main types: internal and external. Internal digital transformation involves the use of digital technologies to improve internal business processes and operations. This might include things like using digital tools to streamline administrative tasks, automate manufacturing processes, or improve supply chain management. Essentially, internal digital transformation is all about making a business more efficient, more productive, and more cost-effective.

On the other hand, external digital transformation refers to the use of digital technologies to improve the way a business interacts with its customers and delivers value to them. This might include things like creating a digital storefront for online sales, developing mobile apps for customer engagement, or implementing social media strategies for brand awareness. Essentially, external digital transformation is all about creating new channels of communication with customers, improving customer experience, and increasing revenue.

While both types of digital transformation are important, they have different goals and require different approaches. Internal digital transformation is focused on streamlining operations and increasing efficiency, while external digital transformation is focused on improving customer experience and increasing revenue. However, the two are often interrelated and can be used in tandem to create a more comprehensive digital transformation strategy.

It's worth noting that the distinction between internal and external digital transformation is not always clear-cut. In many cases, digital technologies can be used to improve both internal processes and customer-facing operations simultaneously. For example, implementing a cloud-based CRM system can improve internal communication and streamline workflows, while also providing valuable data on customer behavior and preferences.

Overall, digital transformation is a complex and multifaceted process that requires careful planning and execution. By understanding the difference between internal and external digital transformation and how they can work together, businesses can develop a comprehensive digital transformation strategy that leverages the full potential of digital technology to drive growth and success.

Now there are also risks associated with not embracing digital transformation. Companies that fail to keep up with technological advancements risk losing market share to competitors who are better equipped to meet customer needs and expectations. Additionally, businesses that do not adapt to changing technologies may find themselves struggling to attract and retain top talent, as employees increasingly prioritize working for companies that offer modern tools and technologies.

Real world examples of the benefits of digital transformation can be found in companies like Uber and Airbnb. Both of these companies disrupted their respective industries by leveraging digital technology to create new business models and offer innovative services. Uber's mobile app transformed the taxi industry by enabling customers to easily request and pay for rides, while Airbnb's platform disrupted the hospitality industry by allowing people to rent out their homes as accommodations.

Digital transformation is a vital component of modern business, offering numerous benefits including increased efficiency, improved customer experience, and competitive advantage. Companies that fail to embrace digital transformation risk falling behind their competitors and losing relevance in an increasingly digital world. By leveraging digital technology, businesses can stay ahead of the curve and deliver value to their customers in new and innovative ways. However, the process of digital transformation can be challenging and comes with its own set of unique obstacles.

One of the most significant challenges of digital transformation is integrating digital technology into existing business processes. Many businesses have established processes and systems in place that may not be compatible with new technologies. This can result in disruption, confusion, and resistance from employees. For example, if a business is implementing a new customer relationship management (CRM) system, employees may resist the change due to the learning curve involved in using the new technology.

Another challenge is the cost associated with digital transformation. Adopting new technologies can be expensive, especially for small businesses with limited budgets. However, the cost of not embracing digital transformation can be even greater. According to a report by Dell Technologies, companies that fail to embrace digital transformation risk falling behind their competitors, with the potential to lose up to $1.6 trillion in revenue by 2025.

Additionally, businesses must have a clear digital transformation strategy to ensure success. A strategy that outlines the goals, objectives, and timeline of the transformation can help businesses stay on track and measure progress. Without a clear strategy, businesses risk wasting time and resources on technologies that do not align with their overall objectives. I've decided to include a universal step-by-step digital transformation strategy on the next page that may be able to be applied to your business. I encourage you to use this strategy and pair it with the other industry specific strategies that we will mention in the next chapter.

Digital transformation is a complex process that involves a complete overhaul of the existing business processes and systems. However, there are certain universal steps that every business can follow to ensure a successful digital transformation.

Step 1: Assess Your Current Capabilities and Identify Areas for Improvement

The first step in any digital transformation strategy is to assess your current capabilities and identify areas for improvement. This involves evaluating your existing business processes, technologies, and workforce skills to determine where digital transformation can make the most impact. Conducting a comprehensive analysis will help you identify the gaps between your current state and the desired future state, which will guide your digital transformation efforts.

Step 2: Set Clear Goals and Objectives

Once you have identified areas for improvement, the next step is to set clear goals and objectives for your digital transformation strategy. These should be specific, measurable, achievable, relevant, and time-bound (SMART) goals that align with your overall business strategy. For example, you may want to improve customer experience, increase operational efficiency, or develop new revenue streams.

Step 3: Develop a Roadmap

With your goals and objectives in place, the next step is to develop a roadmap for achieving them. This involves identifying the specific digital technologies and solutions that will enable you to achieve your goals, and outlining a detailed plan for implementing them. The roadmap should be broken down into manageable phases with clear timelines and milestones.

Step 4: Build the Right Team

Digital transformation requires a diverse set of skills and expertise. It is important to build the right team that can execute your digital transformation strategy effectively. This may involve hiring new talent, upskilling existing employees, and partnering with external service providers or consultants.

Step 5: Implement and Monitor Progress

With your roadmap and team in place, it is time to implement your digital transformation strategy. This involves executing your plan, monitoring progress, and making adjustments as needed. Regular monitoring and reporting will help you track progress against your goals and identify areas that require additional attention.

Step 6: Continuously Improve

Digital transformation is an ongoing process, and businesses must continuously improve and adapt to stay competitive. It is important to keep your digital transformation strategy up-to-date with emerging technologies, market trends, and changing customer needs.

In summary, digital transformation is a complex process that requires a clear strategy, the right team, and a commitment to continuous improvement. By following these universal steps, businesses can successfully navigate the challenges of digital transformation and achieve their goals. According to a survey by IDC, 85% of companies that have adopted a digital transformation strategy reported increased revenue and improved customer satisfaction. Therefore, digital transformation is not just a trend, but a necessity for businesses to remain competitive and thrive in the digital age.

Now to overcome these challenges, businesses must take a proactive approach to digital transformation. This includes identifying the specific areas where technology can enhance operations, prioritizing investments in new technologies, and providing training and support to employees during the transition.

One example of a business successfully implementing digital transformation is Domino's Pizza. In 2009, the company launched an online ordering system, allowing customers to order directly from their website. Since then, Domino's has continued to embrace digital transformation, with innovations such as the Domino's AnyWare platform, which allows customers to order from a variety of devices and platforms, including smartwatches and voice assistants. The company's digital transformation efforts have paid off, with Domino's reporting a 16.9% increase in same-store sales in 2020.

In conclusion, digital transformation is no longer just an option for businesses, but a necessity for survival and growth in today's rapidly changing world. As we have discussed in this chapter, businesses must understand the benefits and risks of digital transformation, as well as the challenges they may face during the process. It is important for companies to develop a clear digital transformation strategy that involves all stakeholders and allows for ongoing evaluation and adjustment.

At Cultivation, we understand the difficulties that companies face in implementing digital transformation strategies, which is why we have developed an artificially intelligent and autonomous platform that automates the entire internal and external process. Our platform is designed to cater to companies of all sizes and budgets, and can help businesses in any industry scale their digital transformation efforts at a cheaper, better and much faster rate.

We believe that by embracing digital transformation and utilizing our platform, businesses can improve their efficiency, competitiveness, and overall success. It is crucial for companies to adapt to the changing business landscape and embrace digital technology in order to thrive in the future.

CHAPTER SEVEN: LET'S STRATEGIZE

The benefits of digital transformation are clear. Organizations that embrace digital technologies are able to streamline operations, improve customer experiences, and stay ahead of the competition. However, the path to digital transformation is not always easy. It requires significant investment in new technology, as well as a willingness to adapt to new ways of working and thinking.

In this chapter, we will explore digital transformation in every industry, from manufacturing and healthcare to retail and financial services. We will examine the specific challenges and opportunities that each industry faces in the digital age, and we will provide practical strategies for driving digital transformation in each sector.

Retail

The retail industry has been rapidly transformed by digital technologies over the past decade. From e-commerce to mobile payments, retailers are using technology to create new shopping experiences and connect with customers in new ways. However, the path to digital transformation is not always easy. It requires significant investment in new technology, as well as a willingness to adapt to new ways of working and thinking. In this section, we will explore a comprehensive digital transformation strategy for the retail industry, including real-world examples of successful implementation.

1. Customer-Centric Approach

The first step in any successful digital transformation strategy for the retail industry is to adopt a customer-centric approach. This means putting the customer at the center of everything you do, from product design to marketing to sales. To achieve this, retailers need to leverage data and analytics to gain insights into customer behavior and preferences. By understanding what customers want, retailers can create personalized experiences that drive loyalty and repeat business.

One example of a retailer that has successfully adopted a customer-centric approach is Sephora. The beauty retailer has invested heavily in technology to create a seamless omnichannel experience for customers. Sephora's Beauty Insider program, for example, provides personalized product recommendations and rewards for loyal customers. The program also enables customers to earn points and redeem them for free products or experiences, creating a sense of exclusivity and loyalty.

2. Streamlined Operations

The second key element of a successful digital transformation strategy for the retail industry is streamlined operations. This means leveraging technology to optimize supply chain management, inventory management, and other operational processes. By streamlining operations, retailers can reduce costs, improve efficiency, and provide faster and more accurate service to customers.

One example of a retailer that has successfully streamlined operations is Walmart. The retail giant has invested heavily in technology to optimize its supply chain management, enabling it to deliver products to stores and customers more quickly and efficiently. Walmart's use of RFID technology, for example, enables it to track inventory in real-time, reducing the risk of out-of-stock situations and ensuring that products are always available for customers.

3. Personalized Marketing

The third key element of a successful digital transformation strategy for the retail industry is personalized marketing. This means using data and analytics to create targeted marketing campaigns that resonate with customers. By delivering personalized marketing messages, retailers can drive customer engagement and loyalty, and ultimately, increase sales.

One example of a retailer that has successfully implemented personalized marketing is Amazon. The e-commerce giant uses data and analytics to create highly targeted marketing campaigns that are tailored to individual customers. For example, Amazon's recommendation engine suggests products based on a customer's purchase history and browsing behavior, creating a personalized shopping experience that drives engagement and loyalty.

4. Seamless Omnichannel Experience

The fourth key element of a successful digital transformation strategy for the retail industry is a seamless omnichannel experience. This means enabling customers to interact with retailers across multiple channels, from in-store to online to mobile. By providing a seamless omnichannel experience, retailers can create a more engaging and convenient shopping experience for customers, driving loyalty and repeat business.

One example of a retailer that has successfully created a seamless omnichannel experience is Nordstrom. The department store chain has invested heavily in technology to enable customers to shop seamlessly across multiple channels. Nordstrom's mobile app, for example, provides personalized product recommendations, enables customers to buy online and pick up in-store, and even allows customers to book appointments with personal stylists.

5. Data-Driven Decision Making

The fifth key element of a successful digital transformation strategy for the retail industry is data-driven decision making. This means leveraging data and analytics to make better business decisions, from product design to marketing to sales. By using data to inform decision making, retailers can reduce risk, optimize performance, and drive innovation.

One example of a retailer that has successfully embraced data-driven decision making is Target. The retailer has invested heavily in data analytics to gain insights into customer behavior and preferences. Target uses this data to inform everything from product design to marketing to store layout. For example, the retailer uses data to identify popular products and trends, enabling it to stock its stores with the products that customers are most likely to buy.

6. Embrace Emerging Technologies

The sixth strategy for digital transformation in the retail industry is to embrace emerging technologies. Retailers need to stay up-to-date with the latest technological advancements to remain competitive. By embracing emerging technologies, retailers can create new shopping experiences and streamline operational processes. Some of the emerging technologies that retailers should consider include:

- Augmented reality (AR) and virtual reality (VR): These technologies can be used to create immersive shopping experiences for customers, allowing them to visualize products in their own homes before making a purchase.

- Artificial intelligence (AI): AI can be used to improve product recommendations, optimize pricing, and personalize marketing messages.

- Internet of Things (IoT): IoT can be used to track inventory in real-time, monitor store traffic, and provide personalized offers to customers based on their location and behavior.

7. Collaborate with Partners

The seventh strategy for digital transformation in the retail industry is to collaborate with partners. Retailers need to work with other companies, such as technology providers and logistics partners, to create a seamless omnichannel experience for customers. By collaborating with partners, retailers can leverage their expertise and resources to create new opportunities for growth.

One example of a retailer that has successfully collaborated with partners is Target. The retailer has partnered with Google to create a voice-activated shopping experience through Google Assistant. Customers can use their Google Home device to add items to their Target cart and place orders through Google Express, creating a seamless shopping experience across multiple channels.

8. Build a Culture of Innovation

The eighth strategy for digital transformation in the retail industry is to build a culture of innovation. Retailers need to encourage experimentation and risk-taking to stay ahead of the competition. By fostering a culture of innovation, retailers can create new products and services that meet the changing needs of customers.

One example of a retailer that has successfully built a culture of innovation is Zara. The fashion retailer is known for its fast fashion approach, with new styles and designs introduced every week. Zara's supply chain is optimized to enable it to design, produce, and deliver new products quickly, allowing the company to respond rapidly to changing fashion trends.

9. Invest in Employee Training

The ninth strategy for digital transformation in the retail industry is to invest in employee training. Retailers need to ensure that their employees have the skills and knowledge required to succeed in a digital environment. By investing in employee training, retailers can improve operational efficiency, drive innovation, and provide better customer service.

One example of a retailer that has successfully invested in employee training is Best Buy. The electronics retailer has implemented a training program called "Geek Squad Academy," which teaches young people technology skills and offers them the opportunity to work in Best Buy stores. By investing in employee training, Best Buy has been able to create a knowledgeable and engaged workforce, which in turn has driven customer satisfaction and loyalty.

10. Secure Data and Protect Customer Privacy

The tenth strategy for digital transformation in the retail industry is to secure data and protect customer privacy. Retailers need to ensure that customer data is protected against cyber threats and that privacy is respected. By securing data and protecting customer privacy, retailers can build trust with their customers, which is essential for long-term success.

One example of a retailer that has successfully secured data and protected customer privacy is Apple. The tech giant has implemented strong security measures to protect customer data, including end-to-end encryption for iMessage and FaceTime. Apple has also taken a strong stance on privacy, with CEO Tim Cook stating that "privacy is a fundamental human right." By securing data and protecting customer privacy, Apple has been able to build a loyal customer base that trusts the company with their personal information.

11. Adopt Agile Methodologies

The eleventh and final strategy for digital transformation in the retail industry is to adopt agile methodologies. Retailers need to be able to respond quickly to changes in the market and customer needs. By adopting agile methodologies, retailers can create a flexible and adaptable organization that can quickly pivot in response to changing circumstances.

One example of a retailer that has successfully adopted agile methodologies is Walmart. The retail giant has implemented an agile methodology called "Walmart Agile," which allows teams to work in a collaborative and iterative way to develop new products and services. By adopting agile methodologies, Walmart has been able to increase innovation and improve time-to-market for new products and services.

Healthcare

Digital transformation has become a critical initiative for the healthcare industry as it strives to improve patient outcomes, reduce costs, and increase efficiency. In this section, we will explore the digital transformation strategies for the healthcare industry, including telemedicine, electronic health records, and artificial intelligence.

1. Telemedicine

Telemedicine, or the delivery of healthcare services remotely through technology, has become increasingly popular in recent years. It has the potential to improve access to healthcare, reduce costs, and increase patient satisfaction. One example of telemedicine in action is Teladoc, a virtual healthcare provider that connects patients with physicians through a mobile app or website. Teladoc provides patients with access to medical professionals 24/7, regardless of their location, and has been proven to reduce the need for emergency room visits and hospitalizations.

Another example of telemedicine is remote patient monitoring, which involves the use of technology to track patient health data in real-time. This can be done through wearable devices that measure vital signs or through sensors placed in the home. Remote patient monitoring has been shown to reduce hospital re-admissions and improve patient outcomes.

2. Electronic Health Records

Electronic health records (EHRs) have become a critical component of the digital transformation of the healthcare industry. EHRs provide healthcare professionals with access to patient medical records in real-time, allowing for more informed decision-making and better coordination of care. One example of EHRs in action is Epic Systems, a healthcare software company that provides electronic health record systems to hospitals and medical practices. Epic's EHR system allows healthcare professionals to access patient records from anywhere, improving efficiency and reducing errors.

3. Artificial Intelligence

Artificial intelligence (AI) has the potential to transform the healthcare industry by improving diagnosis accuracy, predicting outcomes, and enabling personalized medicine. One example of AI in action is IBM Watson Health, which uses AI to analyze vast amounts of medical data to identify patterns and make predictions. Watson Health has been used to improve cancer diagnosis accuracy and predict patient outcomes in clinical trials.

Another example of AI in healthcare is machine learning algorithms that can identify patients at risk for certain conditions, such as heart disease or diabetes. By identifying at-risk patients, healthcare professionals can take preventive measures to reduce the likelihood of serious health complications.

Finance

The finance industry has been undergoing a digital transformation for several years now, driven by technological advancements, changing customer expectations, and new market entrants. In this section, we will explore the digital transformation strategies for the finance industry, including mobile banking, blockchain technology, and data analytics.

1. Mobile Banking

Mobile banking has become an essential component of the digital transformation of the finance industry. With the widespread use of smartphones and tablets, customers expect to be able to access their accounts, make transactions, and manage their finances on-the-go. One example of mobile banking in action is Ally Bank, a digital bank that provides customers with a mobile app that allows them to manage their accounts, deposit checks, and transfer funds. Ally Bank's mobile app has been recognized for its user-friendly interface and advanced security features.

Another example of mobile banking is peer-to-peer payment apps such as Venmo and Cash App. These apps allow users to send and receive money from friends and family instantly, making it easy to split bills and repay debts. Peer-to-peer payment apps have gained popularity among millennials and younger generations who prefer digital payment options over traditional banking methods.

2. Blockchain Technology

Blockchain technology has the potential to revolutionize the finance industry by increasing transparency, reducing costs, and improving security. Blockchain is a decentralized ledger that enables secure, transparent transactions without the need for intermediaries. One example of blockchain technology in action is Ripple, a payment processing platform that uses blockchain to enable real-time, low-cost international money transfers. Ripple's blockchain technology eliminates the need for intermediaries, reducing costs and increasing speed and security.

Another example of blockchain technology in finance is smart contracts, which are self-executing contracts with the terms of the agreement written into code on the blockchain. Smart contracts can automate the processing of transactions, reducing the need for intermediaries and increasing efficiency.

3. Data Analytics

Data analytics is a critical component of the digital transformation of the finance industry, enabling organizations to gain insights into customer behavior, identify new business opportunities, and manage risk. One example of data analytics in action is Capital One, a financial services company that uses data analytics to personalize customer experiences and improve risk management. Capital One's data analytics tools enable the company to analyze customer data in real-time, identify patterns, and provide personalized product recommendations.

Another example of data analytics in finance is credit scoring algorithms that use machine learning to assess creditworthiness. These algorithms can analyze vast amounts of data to identify patterns and make more accurate predictions about an individual's creditworthiness, reducing the likelihood of default and improving risk management.

4. Robotic Process Automation (RPA)

RPA is a technology that automates repetitive, manual tasks, such as data entry and processing. In the finance industry, RPA can help reduce costs and increase efficiency by automating back-office processes. For example, ING Bank uses RPA to automate processes such as account opening and loan processing, reducing the time it takes to complete these tasks and improving the customer experience.

5. Cloud Computing

Cloud computing is a technology that enables organizations to access computing resources, such as servers and storage, over the internet. In the finance industry, cloud computing can help reduce costs, improve scalability, and increase agility. For example, Goldman Sachs recently announced that it is moving its core risk-management platform to the cloud, enabling it to scale quickly and reduce costs.

6. Digital Identity Verification

Digital identity verification is a technology that enables organizations to verify the identity of their customers online, reducing the need for in-person identity checks. In the finance industry, digital identity verification can help improve the customer experience, reduce costs, and enhance security. For example, Jumio is a digital identity verification provider that uses AI and biometrics to verify the identity of customers remotely.

7. Open Banking

Open banking is a regulatory initiative that allows customers to share their financial data with third-party providers, enabling them to access a broader range of services and products. Open banking can help increase competition in the financial services industry, encourage innovation, and improve customer experiences. For example, Monzo, a digital bank in the UK, uses open banking to enable customers to view their accounts from other banks within its mobile app.

8. Chatbots

Chatbots are AI-powered software applications that can interact with customers through messaging platforms, websites, or mobile apps. In the finance industry, chatbots can help automate customer service, answer customer queries, and provide personalized advice. For example, Capital One's Eno chatbot can help customers manage their finances, track their spending, and pay their bills.

9. Cybersecurity

Cybersecurity involves protecting computer systems, networks, and data from unauthorized access, theft, or damage. In the finance industry, cybersecurity is a critical area of concern given the sensitive nature of financial data. Finance organizations need to adopt robust cybersecurity measures to protect customer data and ensure compliance with regulatory requirements. For example, the Financial Services Information Sharing and Analysis Center (FS-ISAC) is a

nonprofit organization that facilitates information sharing and
collaboration on cybersecurity issues among financial institutions.

Manufacturing

Digital transformation is changing the landscape of every industry, and the manufacturing industry is no exception. With advancements in technology such as the Internet of Things (IoT), artificial intelligence (AI), and cloud computing, manufacturers can now optimize their operations and gain insights that were previously impossible. In this chapter, we will discuss a comprehensive digital transformation strategy for the manufacturing industry.

1. Connected Devices

Connected devices, or the Internet of Things (IoT), can be used to gather data from machines and equipment in real-time. This data can then be used to optimize processes and improve product quality. For example, General Electric has developed a system called Predix that can gather data from industrial machines and equipment, analyze it, and provide insights into machine performance and maintenance needs.

2. Artificial Intelligence (AI)

AI can be used to automate processes, improve quality, and optimize operations. For example, AI can be used to detect defects in products, analyze customer feedback, and optimize supply chain management. Amazon is using AI-powered robots to pick and pack products in its warehouses, reducing labor costs and improving efficiency.

3. Additive Manufacturing

Additive manufacturing, also known as 3D printing, is a process that creates products by adding material layer by layer. Additive manufacturing can reduce the cost and time required to produce complex parts, as well as allow for customization of products. For example, GE Aviation is using 3D printing to produce fuel nozzles for its jet engines, reducing production costs and improving performance.

4. Cloud Computing

Cloud computing allows manufacturers to store and access data from anywhere, at any time. This can improve collaboration and communication across teams and locations, as well as allow for more efficient data analysis. For example, Tesla uses cloud computing to collect data from its electric cars and use that data to improve vehicle performance and safety.

5. Robotics

Robots can be used to automate tasks that are dangerous or repetitive for humans. Robots can also work alongside humans, augmenting their capabilities and improving efficiency. For example, Foxconn, a manufacturer of electronics for companies such as Apple and Dell, has installed over 60,000 robots in its factories to perform tasks such as welding and painting.

6. Supply Chain Management

Digital transformation can improve supply chain management by providing real-time data on inventory, orders, and shipments. This can improve efficiency, reduce costs, and improve customer satisfaction. For example, DHL is using IoT sensors to track shipments in real-time, allowing customers to track their packages and receive alerts on their delivery status.

7. Augmented Reality (AR)

AR can be used to provide training and support to workers, as well as improve product design and customer engagement. For example, Boeing uses AR to provide training to its technicians, allowing them to practice complex procedures without the need for actual equipment.

8. Big Data Analytics

Big data analytics can be used to gather insights from large amounts of data, such as customer feedback, machine data, and sales data. This can help manufacturers make more informed decisions, improve product design, and optimize operations. For example, Caterpillar uses big data analytics to analyze machine data and provide insights into maintenance needs, reducing downtime and improving productivity.

9. Cybersecurity

Digital transformation can create new risks, such as data breaches and cyberattacks. Manufacturers need to adopt robust cybersecurity measures to protect their data and operations. For example, Schneider Electric has developed a cybersecurity program that includes risk assessment, vulnerability management, and incident response planning.

10. Predictive Maintenance

Predictive maintenance uses machine learning algorithms to predict when equipment will fail and schedule maintenance accordingly. This can reduce downtime and maintenance costs. For example, Siemens uses predictive maintenance to monitor its wind turbines and schedule maintenance based on usage and weather conditions.

11. Digital Twins

A digital twin is a virtual model of a physical product or process. Manufacturers can use digital twins to simulate and optimize product designs, test different scenarios, and improve performance. For example, NASA used digital twins to test the design of its Mars rover before it was built.

12. Blockchain

Blockchain is a secure, distributed ledger technology that can be used to track and verify transactions. Manufacturers can use blockchain to improve supply chain transparency, reduce the risk of counterfeiting, and ensure compliance with regulations. For example, IBM has developed a blockchain platform called TrustChain that tracks the origin and authenticity of diamonds.

13. Digital Platforms

Digital platforms are online marketplaces that connect manufacturers with suppliers and customers. Manufacturers can use digital platforms to expand their reach, reduce transaction costs, and gain insights into customer preferences. For example, Alibaba has developed a digital

platform called 1688.com that connects Chinese manufacturers with
buyers worldwide.

Transportation

Digital transformation has become a crucial aspect of the transportation industry, as advancements in technology continue to reshape the way people and goods move around the world. By leveraging the power of digital technologies, transportation companies can optimize their operations, enhance safety, and improve the overall customer experience. In this article, we will discuss a comprehensive digital transformation strategy for the transportation industry, highlighting key technologies, real-world examples, and best practices.

1. Embrace the Internet of Things (IoT)

The Internet of Things (IoT) involves the use of connected devices and sensors to gather data and communicate with other devices. In the transportation industry, IoT technologies can be used to monitor and manage vehicles, track shipments, and optimize logistics. For example, FedEx uses IoT sensors to track the location, temperature, and humidity of its shipments, ensuring they arrive in optimal condition.

2. Implement Artificial Intelligence (AI)

Artificial Intelligence (AI) technologies, such as machine learning and natural language processing, can help transportation companies analyze large amounts of data and gain insights into customer behavior and preferences. For example, Uber uses AI algorithms to optimize driver routes and predict demand, improving efficiency and reducing wait times.

3. Enhance Safety with Autonomous Vehicles

Autonomous vehicles have the potential to transform the transportation industry by improving safety, reducing emissions, and increasing efficiency. By using sensors, cameras, and other technologies, autonomous vehicles can navigate roads and highways without human intervention. For example, Waymo, a subsidiary of Alphabet, is developing a self-driving taxi service in Phoenix, Arizona.

4. Utilize Augmented and Virtual Reality

Augmented and Virtual Reality (AR/VR) technologies can be used in the transportation industry to improve training and safety, as well as enhance the customer experience. For example, the logistics company DHL uses AR headsets to guide warehouse workers through complex tasks, improving productivity and reducing errors.

5. Improve Logistics with Digital Twins

Digital twins, which are virtual replicas of physical objects or systems, can be used in the transportation industry to simulate and optimize logistics operations. For example, Maersk Line, the world's largest container shipping company, uses digital twins to analyze data on shipping routes and weather patterns, improving efficiency and reducing fuel consumption.

6. Leverage Blockchain for Secure Transactions

Blockchain is a secure, distributed ledger technology that can be used in the transportation industry to track and verify transactions, as well as improve supply chain transparency. For example, IBM is working with shipping company Maersk to create a blockchain-based platform that will digitize and automate global trade.

7. Implement Real-Time Data Analytics

Real-time data analytics can be used to monitor and analyze transportation operations, allowing companies to make faster and more informed decisions. For example, Delta Airlines uses real-time analytics to monitor flight schedules and predict delays, allowing it to proactively rebook passengers and reduce disruptions.

8. Enhance Customer Experience with Mobile Apps

Mobile apps can be used in the transportation industry to enhance the customer experience by providing real-time information, such as flight schedules, traffic updates, and transportation options. For example, the ride-sharing company Lyft allows customers to track the location of their driver in real-time and rate their experience after the ride.

9. Improve Supply Chain Visibility with RFID

Radio-Frequency Identification (RFID) technology can be used in the transportation industry to improve supply chain visibility and reduce inventory costs. For example, Walmart uses RFID tags to track inventory levels in real-time, allowing it to optimize its supply chain and reduce waste.

10. Integrate Sustainability into Operations

Digital transformation can help transportation companies reduce their environmental impact by optimizing routes, reducing emissions, and improving energy efficiency. For example, the logistics company UPS has implemented a range of sustainability initiatives, including using alternative fuels and optimizing delivery routes to reduce emissions.

11. Foster Collaboration with Industry Partners

Collaboration with industry partners can help transportation companies leverage the latest technologies and best practices to improve operations and customer experience. For example, the airline industry has created the One ID initiative, which aims to use digital technologies to create a seamless, secure travel experience for customers.

12. Invest in Cybersecurity

As transportation companies increasingly rely on digital technologies, cybersecurity has become a critical concern. Investing in robust cybersecurity measures, such as encryption and multi-factor authentication, can help companies protect sensitive data and prevent cyber-attacks. For example, Delta Airlines has implemented a range of cybersecurity measures, including continuous monitoring of its network and threat intelligence sharing with other companies.

13. Develop a Digital Culture

To successfully implement a digital transformation strategy, transportation companies must develop a digital culture that values innovation and continuous learning. This includes investing in employee training and development programs, as well as creating a culture of experimentation and risk-taking. For example, the transportation company Uber encourages employees to experiment with new ideas and technologies through its Hackathon program.

14. Use Predictive Maintenance to Improve Fleet Management

Predictive maintenance involves using data and analytics to identify potential maintenance issues before they occur, improving fleet management and reducing downtime. For example, the transportation company FirstGroup uses predictive maintenance technologies to monitor the health of its bus fleet and optimize maintenance schedules.

15. Leverage Big Data Analytics to Optimize Operations

Big data analytics can be used in the transportation industry to analyze vast amounts of data and gain insights into customer behavior, market trends, and operational performance. For example, the airline industry uses big data analytics to analyze passenger data, optimize pricing, and improve revenue management.

Hospitality

Digital transformation has become a critical aspect of the hospitality industry, as hotels and resorts seek to provide a better customer experience, improve operational efficiency, and gain a competitive advantage in the market. By leveraging technologies such as mobile devices, artificial intelligence (AI), internet of things (IoT), and blockchain, hospitality companies can optimize their operations, reduce costs, and create unique and personalized experiences for their guests. In this section, we will explore a comprehensive digital transformation strategy for the hospitality industry, including specific tactics and real-world examples.

1. Enhance Customer Experience with Mobile Technology

Mobile technology has become ubiquitous in the hospitality industry, providing guests with the convenience and flexibility they desire. Hotels can leverage mobile devices to provide a range of services to their guests, such as mobile check-in, keyless entry, room service orders, and concierge services. For example, the Marriott International hotel chain has implemented mobile check-in and keyless entry at many of its properties, allowing guests to bypass the front desk and go straight to their rooms using their smartphones.

2. Use AI to Personalize Guest Experience

Artificial intelligence (AI) can be used in the hospitality industry to provide personalized experiences to guests, based on their preferences and behavior. AI-powered chatbots can be used to provide instant customer support and answer common questions, while also collecting data on guest preferences and behavior. This data can be used to tailor the guest experience, such as recommending nearby attractions, restaurants, and events based on their interests. For example, Hilton Worldwide has implemented Connie, an AI-powered concierge robot that provides guests with information on hotel amenities, local attractions, and more.

3. Implement IoT to Optimize Operations

The internet of things (IoT) can be used in the hospitality industry to monitor and optimize a range of operations, such as energy usage, room occupancy, and maintenance. Sensors can be installed in guest rooms and common areas to monitor temperature, lighting, and energy usage, allowing hotels to optimize energy efficiency and reduce costs. IoT devices can also be used to monitor room occupancy and schedule cleaning and maintenance tasks, improving operational efficiency and guest satisfaction. For example, the Wynn Las Vegas resort has implemented a smart building system that uses IoT sensors to monitor temperature, lighting, and energy usage, resulting in significant energy savings.

4. Leverage Blockchain for Secure and Transparent Transactions

Blockchain technology can be used in the hospitality industry to provide secure and transparent transactions, such as payment processing, loyalty rewards, and supply chain management. By using blockchain, hotels can improve the security and transparency of their transactions, while also reducing transaction costs and eliminating the need for intermediaries. For example, the hotel booking platform Travala.com uses blockchain technology to provide a secure and transparent payment system, allowing users to pay for bookings using various cryptocurrencies.

5. Use AR/VR to Enhance Guest Experience

Augmented reality (AR) and virtual reality (VR) can be used in the hospitality industry to enhance the guest experience, such as providing virtual tours of hotels and attractions, or showcasing personalized room layouts and amenities. AR/VR can also be used to provide immersive training for employees, improving their skills and performance. For example, the Marriott International hotel chain has implemented a virtual reality training program for its employees, allowing them to practice real-world scenarios in a safe and controlled environment.

6. Implement Contactless Technology for Health and Safety

In the wake of the COVID-19 pandemic, contactless technology has become increasingly important in the hospitality industry, allowing guests to minimize physical contact and maintain social distancing. Hotels can implement contactless technology such as contactless payments, mobile check-in, and virtual room keys to reduce the risk of infection and improve guest safety. For example, the Accor hotel chain has implemented a range of contactless technologies, including mobile check in.

Energy

The energy industry is facing significant challenges as the world moves towards more sustainable and renewable sources of energy. Digital transformation can play a crucial role in helping energy companies to navigate this transition and to remain competitive in a rapidly changing landscape. In this section, we will explore some of the key strategies that energy companies can use to successfully implement digital transformation.

1. Smart Grids

Smart grids are intelligent electrical grids that use data analytics and automation to improve the efficiency, reliability, and sustainability of energy distribution. Smart grids use sensors and monitoring systems to collect data on energy usage and demand, allowing energy companies to optimize their energy production and distribution. For example, the implementation of smart grids in the US has reduced the number of power outages by 30%, saving billions of dollars in economic losses.

2. Internet of Things (IoT)

The Internet of Things (IoT) refers to the network of interconnected devices that are embedded with sensors, software, and connectivity to collect and exchange data. In the energy industry, IoT can be used to monitor energy consumption, optimize energy usage, and improve energy efficiency. For example, energy companies can use IoT-enabled devices to remotely monitor and control energy usage in commercial buildings, reducing energy waste and lowering costs.

3. Artificial Intelligence (AI)

AI can be used to analyze large amounts of data and to identify patterns and trends that are difficult for humans to detect. In the energy industry, AI can be used for a range of applications, from predicting energy demand to optimizing energy production and distribution. For example, GE has developed an AI-powered software platform that can analyze data from wind turbines to predict maintenance needs and to optimize turbine performance.

4. Renewable Energy

The transition to renewable energy sources is a major driver of digital transformation in the energy industry. Renewable energy sources such as solar, wind, and hydroelectric power require advanced digital technologies to optimize their energy production and distribution. For example, energy companies can use data analytics to predict wind and solar energy production and to optimize energy storage and distribution.

5. Blockchain

Blockchain is a decentralized digital ledger that can be used to create secure and transparent transactions. In the energy industry, blockchain can be used for a range of applications, from verifying the authenticity of energy credits to creating secure digital identities for energy producers and consumers. For example, the Brooklyn Microgrid project uses blockchain to create a peer-to-peer energy trading platform, allowing consumers to buy and sell energy from their neighbors.

6. Digital Twins

Digital twins are virtual replicas of physical assets, such as power plants and wind turbines. Digital twins can be used to simulate and optimize the performance of energy assets, allowing energy companies to reduce maintenance costs and improve energy efficiency. For example, Siemens has developed a digital twin platform for wind turbines, allowing energy companies to simulate and optimize turbine performance.

7. Cybersecurity

The digital transformation of the energy industry also presents new cybersecurity risks. Energy companies need to ensure that their digital infrastructure is secure and protected against cyber threats. This can include implementing security protocols, training employees on cybersecurity best practices, and investing in cybersecurity technologies such as intrusion detection systems and encryption.

Construction

The construction industry has traditionally been slow to adopt digital technologies, but the trend towards digitization is now accelerating. Digital transformation is a vital step for construction companies that want to improve productivity, efficiency, and safety on their job sites. This transformation is driven by the adoption of emerging technologies such as Building Information Modelling (BIM), Internet of Things (IoT), robotics, and artificial intelligence. In this section, we will explore a digital transformation strategy for the construction industry, outlining the key technologies and practices that can help construction companies achieve success.

1. Assessment of Current Systems

The first step in any digital transformation strategy is to assess the current systems and technologies that are in use. This step involves analyzing the construction company's current workflows, data management systems, and communication methods to identify areas that could be improved through digital technology.

One area that is ripe for improvement is data management. Construction companies typically generate large volumes of data from multiple sources such as BIM, job site sensors, and drones. This data can be difficult to manage and analyze using traditional methods. Adopting a cloud-based data management system that can process large volumes of data and provide real-time insights can help construction companies to improve decision-making and efficiency.

Another area where digital transformation can have a significant impact is in communication. Job sites can be chaotic, and communication breakdowns can cause delays, safety issues, and rework. Adopting a digital communication system that can provide real-time updates and allow workers to communicate quickly and easily can help to reduce errors and improve safety.

2. Adoption of Building Information Modelling (BIM)

BIM is a powerful technology that enables construction companies to create a digital model of a building or infrastructure project. BIM provides a centralized platform for sharing information and allows stakeholders to collaborate in real-time. This technology can help to improve productivity and efficiency by reducing errors and rework.

Adopting BIM requires a significant investment in technology and training, but the benefits can be substantial. For example, a study by McGraw Hill found that construction companies that used BIM saw an average of 7% reduction in project costs and a 5% reduction in project duration.

3. Integration of Internet of Things (IoT) and Sensors

IoT and sensors can be used to collect real-time data on job sites, such as temperature, humidity, and movement. This data can be analyzed to provide insights into productivity, safety, and equipment usage. IoT and sensors can also be used to automate processes such as material ordering and delivery.

For example, a construction company in the UK used sensors to track the usage of construction equipment. The company was able to optimize equipment usage, resulting in a 50% reduction in rental costs.

4. Adoption of Robotics and Automation

Robotics and automation can be used to perform repetitive and dangerous tasks, such as bricklaying and concrete pouring. These technologies can help to reduce the risk of injury to workers and improve productivity.

For example, a construction company in Japan used a robot to lay 1,000 bricks in one hour, a task that would have taken a human worker 3 days to complete.

5. Implementation of Artificial Intelligence (AI)

AI can be used to analyze large volumes of data and provide insights into productivity, safety, and equipment usage. AI can also be used to predict potential problems and provide recommendations for improving efficiency and safety.

For example, a construction company in the US used AI to analyze data from job site sensors and predict safety hazards. The company was able to reduce safety incidents by 35%.

6. Training and Education

Digital transformation requires a significant investment in technology and training. Construction companies must ensure that their workers are trained on new technologies and processes to maximize the benefits of digital transformation. This investment in training and education can also help to improve worker retention and attract new talent.

7. Predictive Analytics

Predictive analytics is a technology that uses data, statistical algorithms, and machine learning techniques to identify the likelihood of future outcomes based on historical data. In the construction industry, this technology can be used to predict project delays, cost overruns, and other potential issues.

By analyzing data from past construction projects, predictive analytics can identify patterns and trends that can help project managers make better decisions and avoid potential problems. This can lead to more efficient and cost-effective construction projects, as well as improved safety and quality.

8. Virtual and Augmented Reality

Virtual and augmented reality technologies are also being used in the construction industry to improve planning, design, and communication. These technologies allow construction professionals to create 3D models of buildings and other structures, which can be used to visualize the project and identify potential issues before construction begins.

In addition, virtual and augmented reality can be used to train workers and improve safety. For example, workers can use these technologies to practice tasks in a simulated environment before performing them on a real construction site.

9. Blockchain

Blockchain is a distributed ledger technology that allows multiple parties to share and verify data in a secure and transparent way. In the construction industry, blockchain can be used to improve supply chain management and reduce the risk of fraud and errors.

By using blockchain to track the movement of materials and products throughout the supply chain, construction companies can ensure that materials are delivered on time and that they meet quality standards. In addition, blockchain can be used to verify the authenticity of documents and contracts, reducing the risk of fraud and disputes.

Mining

The mining industry is an essential sector that provides raw materials for a wide range of industries such as construction, manufacturing, energy, and technology. The digital transformation of the mining industry can enhance operational efficiency, reduce costs, increase safety, and improve sustainability. The mining industry has already started to leverage digital technologies to optimize their operations, including predictive maintenance, automation, and advanced analytics.

In this section, we will discuss a digital transformation strategy for the mining industry. We will focus on various aspects of the mining value chain and how digital technologies can be leveraged to drive value. We will also provide real-world examples of companies that have successfully implemented digital transformation initiatives in the mining industry.

1. Exploration and Drilling

Exploration and drilling are the initial stages of the mining value chain. Digital technologies can significantly improve the efficiency and accuracy of these processes. For example, drones can be used for geological mapping, identifying ore bodies, and monitoring the progress of exploration activities. This can save time, reduce costs, and improve safety. Similarly, digital technologies can be used for drilling optimization by analyzing drilling data to optimize drill patterns, drilling parameters, and target selection.

Real-world example: Rio Tinto has implemented an automated drilling system that uses data analytics and artificial intelligence to optimize drilling parameters and drill hole patterns. The system has reduced the time required for drilling by 50%, increased the accuracy of drilling, and improved safety.

2. Mining Operations

Mining operations involve the extraction of minerals from the earth. Digital technologies can improve operational efficiency, reduce costs, and improve safety in mining operations. For example, automation can be used for loading and hauling operations, which can reduce labor costs, improve equipment utilization, and increase safety. Similarly, advanced analytics can be used for predictive maintenance, which can reduce downtime and maintenance costs.

Real-world example: BHP has implemented an autonomous haulage system at its mine sites, which has in fact improved operational efficiency, reduced labor costs, and increased safety.

3. Mineral Processing

Mineral processing involves the extraction of minerals from ore and refining them into marketable products. Digital technologies can improve the efficiency of mineral processing operations by optimizing process parameters, reducing energy consumption, and improving product quality. For example, advanced process controls can be used for optimizing mineral processing operations, and machine learning can be used for identifying process anomalies and optimizing process parameters.

Real-world example: Vale has implemented a machine learning-based system for mineral processing optimization, which has improved process efficiency, reduced energy consumption, and improved product quality.

4. Supply Chain Management

Supply chain management is a critical aspect of the mining industry, as it involves the transportation of minerals from mine sites to end-users. Digital technologies can improve supply chain efficiency, reduce costs, and increase transparency. For example, blockchain technology can be used for tracking minerals from mine to market, which can increase transparency and traceability. Similarly, digital twins can be used for simulating supply chain operations and optimizing logistics.

Real-world example: IBM and MineHub have developed a blockchain-based platform for tracking minerals from mine to market, which has increased transparency and traceability in the supply chain.

5. Sustainability

Sustainability is becoming increasingly important for the mining industry, as stakeholders demand more responsible mining practices. Digital technologies can play a significant role in improving sustainability in the mining industry. For example, advanced analytics can be used for optimizing energy consumption and reducing carbon emissions. Similarly, digital technologies can be used for water management, waste management, and reclamation.

Real-world example: Anglo American has implemented a digital water management system that uses advanced analytics and sensors to optimize water consumption, reduce water waste, and improve water quality.

Agriculture

Agriculture is the backbone of most economies, and in today's fast-paced world, it is imperative to leverage technology to boost productivity and efficiency while reducing environmental impact. In this last section, we will discuss digital transformation strategies for the agriculture industry, including precision farming, smart agriculture, data-driven decision making, and sustainable agriculture practices.

1. Precision Farming

Precision farming, also known as site-specific crop management, uses various technologies to provide farmers with precise information about their crops. This information includes soil health, moisture content, and pest infestations, among others. The data gathered can then be used to make informed decisions about planting, fertilization, irrigation, and pest control. This results in better crop yields, improved resource management, and reduced costs.

One real-world example of precision farming is John Deere's Precision Ag Solutions. John Deere offers several solutions, including AutoTrac™, which is an automated steering system that allows farmers to steer their tractors and other farm equipment with accuracy, and ExactApply™, which provides precise application of liquid fertilizer and crop protection products. This precision ensures that farmers apply the right amount of fertilizer and crop protection products at the right time and in the right place, leading to better crop yields and cost savings.

2. Smart Agriculture

Smart agriculture involves the use of IoT (Internet of Things) devices, sensors, and other technologies to monitor and manage crops in real-time. This enables farmers to make data-driven decisions about crop growth, soil moisture, and nutrient levels. The real-time data also enables farmers to detect potential problems before they become major issues. This approach leads to more efficient resource usage, reduced costs, and increased yields.

One example of smart agriculture is the use of drones in farming. Drones equipped with sensors can gather data about crops, including moisture content, plant health, and temperature. This data can then be used to make decisions about planting, fertilization, and pest control. Drones can also be used to map fields, monitor irrigation, and identify areas where soil needs to be amended. The use of drones enables farmers to gather data in a timely and cost-effective manner, leading to better crop yields and increased profits.

3. Data-Driven Decision Making

Data-driven decision making involves using data analytics to make informed decisions about farming practices. The data used includes information about weather patterns, soil health, crop yields, and pest infestations, among others. Data analytics can also help identify trends and patterns that may not be visible to the naked eye, enabling farmers to make more informed decisions.

One example of data-driven decision making is Climate Corporation's Climate FieldView™. Climate FieldView™ uses data analytics to provide farmers with real-time information about their crops, including yield potential, soil health, and pest pressure. This information enables farmers to make more informed decisions about planting, fertilization, and pest control. The data collected by Climate FieldView™ can also be used to identify patterns and trends, enabling farmers to make more informed decisions about future planting cycles.

4. Sustainable Agriculture Practices

Sustainable agriculture practices involve using technologies that promote the conservation of natural resources, reduce environmental impact, and improve the livelihoods of farmers. These practices include conservation tillage, cover cropping, and crop rotation. The use of sustainable agriculture practices ensures that farms remain productive while minimizing the impact on the environment.

One real-world example of sustainable agriculture practices is the use of precision irrigation systems. These systems use sensors and other technologies to monitor soil moisture levels and deliver water precisely where it is needed. This approach reduces water waste, promotes water conservation, and reduces the impact on the environment. The use of precision irrigation systems also leads to cost savings, as farmers no longer need to over-irrigate their crops to compensate for potential

As we come to the end of our discussion on industry-specific digital transformation strategies, it is evident that digital transformation is becoming increasingly important across all industries. With technology evolving at a rapid pace, businesses cannot afford to remain stagnant and must embrace digital transformation to remain competitive in their respective markets.

We have seen how each industry has its unique challenges and opportunities, and digital transformation strategies must be tailored to address those specific needs. For example, the healthcare industry can benefit greatly from the use of telemedicine and AI-powered diagnostics, while the hospitality industry can leverage data analytics and mobile technologies to enhance customer experiences.

It is crucial for businesses to understand that digital transformation is not a one-time process but rather a continuous journey. It requires a cultural shift towards embracing innovation, collaboration, and agility to adapt to changing market conditions and evolving customer needs.

The success of a digital transformation strategy also depends on effective leadership and a strong focus on change management. Leaders must inspire their teams to embrace the digital journey and provide them with the necessary resources and training to succeed.

In conclusion, digital transformation is no longer an option but a necessity for businesses across all industries. By adopting industry-specific digital transformation strategies, businesses can unlock new opportunities, improve operational efficiency, enhance customer experiences, and ultimately achieve long-term growth and success.

CHAPTER EIGHT: THE IMPACT

Digital transformation has had a significant impact on society, with the potential to change the way we work, live and interact with each other. The rapid pace of technological innovation has led to an increase in automation and artificial intelligence (AI), which has the potential to displace jobs and alter the way we think about work.

The impact of digital transformation on the workforce cannot be overstated. Many industries, from manufacturing to healthcare, have undergone significant changes due to technological advancements. The mining industry, for example, has implemented autonomous vehicles and drones to improve efficiency and safety. In the transportation industry, autonomous vehicles and drones are also being used to improve efficiency and reduce costs.

However, the implementation of these technologies has the potential to displace jobs, leading to concerns about job security and unemployment. In the mining industry, for example, the use of autonomous vehicles has led to the displacement of truck drivers. In the hospitality industry, self-check-in kiosks and mobile ordering systems have reduced the need for human staff. These are just a few examples of how digital transformation has disrupted traditional industries and led to job displacement.

Despite these concerns, digital transformation also presents opportunities for retraining and upskilling. As the job market changes, workers will need to adapt to new technologies and learn new skills. In the transportation industry, for example, workers can be trained to operate and maintain autonomous vehicles. In the hospitality industry, workers can be trained to provide high-level customer service and use new technologies to enhance the guest experience.

To make the most of these opportunities, it is essential to invest in education and training programs that prepare workers for the changing job market. This includes providing access to online training resources and partnering with educational institutions to develop curriculum that aligns with the skills required for the digital economy. This book could serve as a great resource.

In addition, businesses must embrace a culture of lifelong learning to help workers stay up-to-date with the latest technological advancements. This includes providing opportunities for on-the-job training, mentoring, and continuing education.

Governments also have a role to play in supporting workers during the transition to a digital economy. This can include providing access to job training and education programs, offering tax incentives to businesses that invest in workforce development, and providing financial support to workers who have been displaced due to job automation.

Not to mention, the impact of digital transformation on privacy and security is an important consideration for individuals and organizations alike. As more data is collected and processed, there is a greater risk of data breaches and cyberattacks. This can lead to the exposure of sensitive information, including personal and financial data, and can result in financial loss, reputational damage, and legal liability.

One of the key risks of digital transformation is the increased exposure to cyber threats. This can include malware attacks, phishing scams, and other forms of cybercrime. The potential consequences of a successful cyberattack can be severe, including financial loss, theft of intellectual property, and reputational damage. In addition to the direct financial costs associated with these incidents, there may also be indirect costs, such as lost productivity, legal fees, and the need for increased cybersecurity measures.

To mitigate these risks, it is important for organizations to take a proactive approach to cybersecurity. This may involve implementing strong encryption protocols, using multi-factor authentication, and regularly testing systems for vulnerabilities. It is also important to have a comprehensive incident response plan in place to ensure a timely and effective response in the event of a cyberattack.

In addition to cybersecurity risks, there are also concerns around data privacy. As more data is collected and processed, individuals may be at risk of having their personal information compromised. This can include sensitive data such as medical records, financial information, and other personally identifiable information.

To address these concerns, governments and regulatory bodies around the world have implemented strong data privacy laws and regulations. For example, the European Union's General Data Protection Regulation (GDPR) provides a comprehensive framework for data privacy, including requirements for informed consent, data minimization, and the right to be forgotten. Similarly, in the United States, the California Consumer Privacy Act (CCPA) provides a range of data privacy protections for California residents.

In addition to these regulatory measures, organizations must also take steps to protect the privacy of their customers and employees. This may involve implementing data encryption and anonymization protocols, limiting data collection and retention, and providing clear and transparent privacy policies.

Lastly, the impact of digital transformation is not only limited to the economic and societal aspects, but also extends to the environment. As the world becomes increasingly digital, there is a growing concern about the impact of digital technology on the environment. This is mainly due to the high energy consumption associated with the production, use and disposal of digital devices and services.

The influence of digital transformation on the environment is significant, particularly in terms of energy consumption and carbon emissions. The energy consumption associated with the production, use and disposal of digital devices and services is increasing rapidly, with estimates suggesting that the digital technology sector accounts for up to 2% of global carbon emissions. The increasing use of data centers, cloud computing, and other digital infrastructure has also led to a significant increase in energy consumption. In fact, data centers alone are estimated to consume 1% of global electricity and produce 0.3% of global carbon emissions.

However, digital technology also has the potential to reduce environmental impact in various ways. For instance, digital technology can enable the use of renewable energy sources such as solar and wind power. In addition, digital technology can help optimize energy consumption by monitoring and controlling energy use in buildings and factories. Furthermore, digital technology can reduce the need for travel by enabling virtual meetings, remote work, and e-commerce.

Despite the potential benefits of using digital technology to reduce environmental impact, there are also potential drawbacks that need to be considered. For instance, the production of digital devices such as smartphones, laptops, and servers require the mining of rare earth minerals and other raw materials, which can lead to environmental damage and social issues such as forced labor and human rights violations. Moreover, the disposal of digital devices also poses significant environmental risks, as e-waste contains hazardous materials such as lead, mercury and cadmium, which can pollute soil and water and harm human health.

To ensure that the benefits of digital transformation outweigh its potential negative impact on the environment, it is important to adopt sustainable practices in the development and use of digital technology. This includes adopting energy-efficient designs, using renewable energy sources, reducing e-waste by promoting recycling and repair, and adopting circular economy principles that promote the reuse of materials and reduce waste.

In addition, there is a need for policies and regulations that promote sustainable digital practices. Governments can play a crucial role in promoting sustainable practices by implementing energy efficiency standards for digital devices, promoting the use of renewable energy, and incentivizing the development of sustainable digital infrastructure. Moreover, companies can also take steps to reduce their environmental impact by adopting sustainable practices in their supply chains, promoting the circular economy, and investing in sustainable digital infrastructure.

In conclusion, digital transformation is rapidly shaping every aspect of our society. From our personal lives to the business world and beyond, digital technology is changing the way we live, work, and interact with each other. It is bringing new opportunities and challenges that we must face as individuals, organizations, and governments.

The impact of digital transformation on the workforce is significant. While it presents new opportunities for automation and efficiency, it also poses a potential threat to jobs and job security. As a society, we must find ways to retrain and upskill workers to adapt to the changing job market and ensure that no one is left behind.

The impact of digital transformation on privacy and security is also a concern. As we become more reliant on digital technology, the potential risks of data breaches and cyberattacks increase. There is a need for strong data privacy laws and security measures to protect our personal and sensitive information.

Lastly, the impact of digital transformation on the environment is an area that requires more attention. While digital technology can offer benefits to reduce environmental impact, such as reducing paper usage or optimizing energy consumption, it also has its drawbacks, including the production of e-waste and the potential for increased energy consumption. As a society, we must prioritize sustainable practices in the development and use of digital technology to minimize its impact on the environment.

Overall, digital transformation is a complex and evolving phenomenon that requires a multi-faceted approach. We must embrace the opportunities it brings while addressing the challenges it presents. As we continue to integrate digital technology into every aspect of our society, we must ensure that we do so in a way that is ethical, responsible, and sustainable. By doing so, we can harness the full potential of digital transformation to create a better future for all.

CHAPTER NINE: A NEW WORLD

As we bring this book to a close, let's take a moment to reflect on what we've covered so far. We started with an exploration of what digital transformation means, its history and why it's important in today's world. We then delved into futurology, technology in modern business and specific strategies for various industries, including retail, healthcare, finance, manufacturing, transportation, hospitality, energy, construction, and mining.

Throughout this journey, we discussed the impact of digital transformation on the workforce and the need for retraining and upskilling to adapt to the changing job market. We also highlighted the potential risks of job displacement due to automation and artificial intelligence.

Another key point we covered was the impact of digital transformation on privacy and security. We discussed the need for strong data privacy laws and security measures to prevent data breaches and cyberattacks.

We then explored the potential benefits and drawbacks of using digital technology to reduce environmental impact, and the need for sustainable practices in the development and use of digital technology.

Looking to the future, it's clear that digital transformation will continue to play a major role in shaping society. We can expect to see ongoing research and development in the field as new technologies emerge and existing ones are refined.

To fully embrace digital transformation, businesses and individuals must be willing to embrace change and adapt to new technologies. This means investing in the right tools and resources, as well as fostering a culture of innovation and collaboration.

Ultimately, our hope is that this book has provided valuable insights and actionable strategies for anyone looking to navigate the complex world of digital transformation. By embracing the potential of this powerful force, we can build a better future for ourselves and our communities.

We encourage you to navigate to our website, heycultivation.com, where you can find more information on our company and how we're developing an autonomous & artificially intelligent digital transformation platform.

Over 90% of businesses will be digitally disrupted, yet less than half, 43% to be exact, are truly prepared for the technological shift. Even worse, less than 15% of employees feel as if they have the digital skillset to carry their companies through this change.

Cultivation can improve those rates tremendously...

Our mission is to lower the rate of business failure and prepare more individuals and companies for the next wave of digitization.

We plan on doing this by creating a level playing field for all business sizes by providing small and medium-sized businesses with the knowledge, manpower, and technology to scale at a cheaper, better, and faster rate...all from one dashboard.

If you are interested in joining a powerful community and want to further the conversation on digital transformation and technology as a whole, we encourage you to join our Discord server at discord.gg/TaD8Mg9t!

Keep up to date with our company by following us on all social media platforms, @heycultivation.

Thank you for taking this journey with us.

BIBLIOGRAPHY

1. Adner, R. (2017). Ecosystem as structure: An actionable construct for strategy. Journal of Management, 43(1), 39-58.
2. Agrawal, A., Gans, J., & Goldfarb, A. (2018). Prediction machines: The simple economics of artificial intelligence. Harvard Business Press.
3. Basu, S., Dey, T. K., & Mukherjee, A. (2018). Smart manufacturing: recent trends and research direction. Journal of Cleaner Production, 197, 989-1004.
4. Bejma, J., & Dosch, C. (2020). Industry 4.0 as an opportunity for sustainable development. Energies, 13(17), 4365.
5. Berman, S. J., & Hagan, P. (2018). Digital transformation: A roadmap for billion-dollar organizations. Routledge.
6. Brynjolfsson, E., & McAfee, A. (2014). The second machine age: Work, progress, and prosperity in a time of brilliant technologies. WW Norton & Company.
7. Bughin, J., Correa, M., & Kogan, J. (2018). The case for digital reinvention. McKinsey Quarterly, 1-12.
8. Bughin, J., Hazan, E., & Ramaswamy, S. (2018). The impact of AI in business: Widespread job losses are not a certainty. Harvard Business Review, 15, 1-6.
9. Chang, H. H., & Chen, S. W. (2019). Digital transformation strategy for the construction industry: A perspective from the Internet of Things. Sustainability, 11(13), 3609.
10. Chui, M., Manyika, J., & Bughin, J. (2016). A future that works: Automation, employment, and productivity. McKinsey Global Institute.
11. Cusumano, M. A. (2018). Technology strategy and management: the example of Platform-based ecosystems. MIS Quarterly, 42(1), 1-24.
12. de Reuver, M., Bouwman, H., & Haaker, T. (2018). Digital transformation: A review of the field and agenda for future research. Journal of Information Technology, 33(2), 129-156.
13. Deloitte. (2019). Industry 4.0: At the intersection of readiness and responsibility. Retrieved from https://www2.deloitte.com/content/dam/Deloitte/in/Docum

ents/manufacturing/in-manufacturing-industry-4-0-noexp.pdf

14. Demirkan, H., Spohrer, J., Welser, J. J., & Watson, R. T. (2016). Building blocks for a theory of service systems. Journal of Service Research, 19(4), 423-428.

15. Dubois, A., & Gadde, L. E. (2019). Digital servitization: Disentangling the hype. Journal of Business Research, 98, 365-376.

16. European Commission. (2019). Digital transformation scoreboard 2019. Retrieved from https://ec.europa.eu/growth/tools-databases/dem/monitor/sites/default/files/DTM_2019_web.pdf

17. Gartner. (2020). Top strategic technology trends for 2020. Retrieved from https://www.gartner.com/smarterwithgartner/gartner-top-10-strategic-technology-trends-for-2020/

18. Gawer, A., & Cusumano, M. A. (2014). Industry platforms and ecosystem innovation. Journal of Product Innovation Management, 31(3), 417-433.

19. Graetz, G., Michaels, G., & Michaels, M. (2018). Robots at work. Review of Economics and Statistics, 100(5), 753-768.

20. Grant, R. M. (2016). Contemporary strategy analysis: Text and cases edition. John Wiley & Sons.

21. Gupta, A., & George, B. (2016). Toward the development of a big data analytics capability. Information & Management, 53(8), 1049-1064.

22. Haarstad, H., & Flak, L. S. (2018). The role of digitalization in the transformation of the energy sector. Energy Research & Social Science, 39, 263-270.

23. Harnesk, D., & Lundqvist, M. (2018). Industry 4.0 readiness in Swedish manufacturing SMEs. Journal of Manufacturing Technology Management, 29(4), 658-675.

24. Heikkilä, J., Heikkilä, T., & Småros, J. (2019). Digital twins in the manufacturing industry: A literature review. Journal of Manufacturing Systems, 51, 121-139.

25. KPMG. (2019). The future of healthcare: Enabling digital transformation with AI. KPMG International.

26. McKinsey & Company. (2019). Industry 4.0: How to navigate digitization of the manufacturing sector. McKinsey & Company.

27. Mims, C. (2017). Is the world ready for robot lawyers? MIT Technology Review.

28. Mollick, E. (2018). The dynamics of crowdfunding: An exploratory study. Journal of Business Venturing, 33(1), 1-17.

29. National Institute of Standards and Technology. (2019). Securing the internet of things. National Institute of Standards and Technology.

30. Rifkin, J. (2011). The third industrial revolution: How lateral power is transforming energy, the economy, and the world. Macmillan.

31. Schwab, K. (2016). The fourth industrial revolution. World Economic Forum.

32. Tran, V. T., & Geertman, S. (2018). Smart cities: Definitions, dimensions, performance, and initiatives. Journal of Urban Technology, 25(2), 1-21.

33. Weill, P., & Woerner, S. L. (2018). What's your digital business model? Harvard Business Review, 96(3), 104-113.

34. World Economic Forum. (2019). The global risks report. World Economic Forum.